AIR AND WASTE MANAGEMENT

HOW TO ORDER THIS BOOK

BY PHONE: 800-233-9936 or 717-291-5609, 8AM–5PM Eastern Time

BY FAX 717-295-4538

BY MAIL: Order Department
Technomic Publishing Company, Inc.
851 New Holland Avenue, Box 3535
Lancaster, PA 17604, U.S.A.

BY CREDIT CARD: American Express, VISA, MasterCard

AIR and WASTE MANAGEMENT

A Laboratory/Field Handbook

Howard E. Hesketh, Ph.D., PE, DEE

Professor of Air and Hazardous Waste Management Engineering
College of Engineering
Southern Illinois University at Carbondale
Carbondale, IL

CRC Press
Taylor & Francis Group
Boca Raton London New York

CRC Press is an imprint of the
Taylor & Francis Group, an **informa** business

CRC Press
Taylor & Francis Group
6000 Broken Sound Parkway NW, Suite 300
Boca Raton, FL 33487-2742

First issued in hardback 2018

ISBN-13: 978-1-56676-111-6 (pbk)
ISBN-13: 978-1-138-45951-9 (hbk)

Main entry under title:
Air and Waste Management: A Laboratory/Field Handbook

A Technomic Publishing Company book
Bibliography: p.
Includes index p. 253

Library of Congress Catalog Card No. 94-61075
ISBN No. 1-56676-111-5

Visit the Taylor & Francis Web site at
http://www.taylorandfrancis.com

and the CRC Press Web site at
http://www.crcpress.com

Table of Contents

Preface

THIS book *Air and Waste Management: A Laboratory/Field Handbook* is a compendium of work by many individuals. Much of the original writing was carried out by the Federal Environmental Protection Agency, by various state agencies, by organizations, by individuals and by universities, as well as by the former Federal Public Health Service Department. The materials used for this manual were contributed by my associates who are named on the following list. Many thanks to each. Other material was contributed which I could not use here, and I thank those persons for their consideration.

Hopefully, this book will serve a useful function in training and giving experience to environmental scientists at all levels of experience (high school, college and continuing education). Included in this book is explanatory materials, exercises and experiments. These are intended as training guides. The readers will find it possible to use simple equipment and naturally occurring events to construct some of the needed equipment but may also find it necessary to use commercially available equipment with some of the procedures. *Federal Register* notations are applicable in a number of examples, and these are referenced.

The Southern Illinois University Department of Mechanical Engineering and Energy Processes provided release time for this project and I appreciate this very much. Also, special thanks go to the departmental word processing secretary, Amy Dunn, for preparing the manuscript; the departmental office manager, Judi Cockrum, for arranging schedules so this work could be completed; the departmental student worker, Leslie Reese, for preparing figures and copies; and to my engineering student, Scott Lane, for helping proofread the manuscript.

HOWARD E. HESKETH, PHD, PE, DEE
Carbondale, IL

List of Contributors

HAROLD M. COTA, PHD, PE, DEE, Environmental Engineering, California Polytechnic State University, San Luis Obispo, CA 93407

CHARLES D. PRATT, PE, DEE, Air Pollution Training Branch, USEPA, Office of Air Quality Planning & Standards, Research Triangle Park, NC 27711

LEO STANDER, PHD, Air Pollution Training Branch, USEPA, Office of Air Quality Planning & Standards, Research Triangle Park, NC 27711

FRANK W. SHERMAN, PE, Environmental Engineer, 3905 W. Bluffs Rd., Springfield, IL 62707

MIKE RUBY, PHD, PE, DEE, Envirometrics, 4803 Fremont Ave., North, Seattle, WA 98103

RICHARD YOUNG, PE, REM, Pollution Engineering, Cahners Publishing Co., 1350 Touhy Ave., Des Plaines, IL 40018

DAVID G. STEPHEN, Senior Engineering Advisor, Risk Reduction Engineering Lab, USEPA, Cincinnati, OH 45268

WILLIAM H. PROKOP, CE, DEE, Prokop Enviro Consulting, P.O. Box 602, Deerfield, IL 60015

HOWARD OLSON, PHD, Colorado Dept. of Health, Air Pollution Tech. Service, Denver, CO 80220

JOZEF PASTUSZKA, PHD, Instytut Ochrany Srodowiska, Environ. Pollution Abatement Centre 40-832, Katowice, Poland,

THOMAS J. OVERCAMP, PHD, PE, DEE, College of Engineering, Clemson University, Environ. Research Lab, Clemson Research Park, SC 29634-0919

TIM C. KEENER, PHD, College of Engineering, University of Cincinnati, Cincinnati, OH 45221-0071

JEFFREY C. SMITH, Executive Director, Industrial Gas Cleaning Institute, 1707 L St., NW, Suite 570, Washington, DC 20036

FRANK L. CROSS, PE, DEE, HLA/CTA, 4763 S. Conway Rd., Orlando, FL 32812

CHAPTER 1

Documentation and General Calibration

INTRODUCTION

To be of value, the procedures and findings of any study must be properly documented, and equipment must be calibrated. This includes properly recording the procedures and data and effectively writing a report. Research procedures and data should always be recorded in a bound notebook with numbered pages. Each page should be dated and signed or initialed. Strip-chart and similar loose data may be taped onto notebook pages. List all equipment used by name and serial number, and identify each on a flow diagram. Record calibration values and operation procedures.

Provide a complete paper trail with your notebook so that legal verification is possible for patents and other needs. For example, sample analysis results data will be entered when received (usually sometime later), but be certain to refer back to the page showing when samples were taken.

REPORT PREPARATION

The engineer or scientist in pursuit of his profession is generally given an objective to achieve. As soon as he begins organizing his thoughts, he is unconsciously preparing a final report on the subject! After all, a report is nothing more than the effective communication of his thought processes. Following this logic, we can list some suggestions to help prepare a report. Notice how closely the following list simulates the logical thought pattern a researcher will use in attacking any problem.

(1) Define the objectives clearly.
(2) Review how the objectives were achieved.
(3) Tabulate the experimental data.
(4) Describe what additional quantities are to be calculated from the experimental data; calculate and tabulate the results of the calculations.

(5) Decide what is the best way of presenting these results: analytical equations, graphs or tables. Prepare a clear presentation.
(6) Analyze your results from Step 5 and prepare a clear presentation, examine their meaning and compare the results with applicable theory or similar results in the literature.
(7) *Scope* your results! What use can be made of your results? What effect does your result have on the design of a piece of equipment? If a design is part of your objective, is the design optimized? What quantitative comparison can you make between your data and any applicable theory? Have you chosen the *most* basic method of comparing your results in order to be able to extrapolate to other operating conditions? What limits would you place on extrapolating your results? How did you arrive at these limits?
(8) Based on Steps 6 and 7, decide on the conclusions and recommendations.
(9) *Outline the report*—First list the sections that will be required to effectively communicate your results and conclusions (hopefully for any level of reader intelligence). Then use a *sentence* outline of topics to be covered in each section, indicating the location of figures, tables, graphs and appendices.
(10) Write the report following the outline closely.
(11) If you are working with a group of students, edit and revise the report as a group, using constructive criticism between members to polish the final product. Be sure the report is:
   - written in good English
   - logical
   - simple and direct
   - technically valid

A relatively simple "communication formula" can be applied to *both* the overall report organization and the organization of *each* section in the report. It is arranged in four successive steps:

(1) Ah-ha (AH)
(2) What Brought That Up (WBTU)
(3) For Instance (FI)
(4) Now Who'll Take Two? (NWTT)

The Ah-ha is an eyebrow-raising phrase related to the subject which serves to awaken the reader and whet his appetite to read on. The WBTU then bridges the gap between the Ah-ha and the subject at hand—it prepares the reader for the basis of your argument or proposal. The For Instance section contains examples or data or theory upon which your argument is based. The FI section can be lengthened or shortened to suit the occasion. However, in report writing, only the examples upon which your results are based should be included–secondary arguments should be deleted for brevity. The Now Who'll Take Two is your close. It presents your argument in the form of a result, it moves the reader to take action, it is the sole reason for all the words which precede it.

Without an emphatic close, the NWTT, there is no reason to present examples or introduce the reader to the subject matter. Therefore, the proper mental organization in a report (or report section) begins with the NWTT and works backward through the FI's, WBTU and AH. The proper test of any report, report section, paragraph, or sentence is whether or not there is NWTT directly attached to *any* item in a report. If it doesn't prove a point, it doesn't belong in the report!

A professional report must have a professional appearance. Lab reports can be handwritten if your handwriting is neat and you do not know how to type (you should take the time to learn). However, a typewritten lab report is preferred, and a Master's Thesis/Project Report must be typewritten. A fresh, black ribbon should be used. The body of the text must be double-spaced. Quotations, footnotes, bibliographies and table/figure captions may be single-spaced. Note that detailed requirements on margins, etc., for Graduate Schools' Theses and Dissertations may need to be followed.

SECTIONS OF THE REPORT

The sections of a report are listed below in order of their placement in a Formal Report.

(1) Title Page
(2) Table of Contents
(3) Abstract
(4) Introduction } 1 or 2
(5) Theory }
(6) Equipment and Materials
(8) Results
(9) Discussion of Results } 1, 2 or 3 sections
(10) Conclusions }
(11) Recommendations
(12) Nomenclature (optional)
(13) References
(14) Appendices

This organization presupposes sufficient material in each section to warrant presentation as a separate entity. Obviously, if only one basic equation is used in the entire report, there would be no need for either a Theory or Nomenclature section. Alternate organizations could include the Abstract presented prior to the Table of Contents, the Conclusions and Recommendation appearing immediately after the Introduction and the Nomenclature on the last page of the report as a foldout for easy reference during reading. The list of report sections does not include the so-called Purpose section, since in this organization that function can best be covered in the Introduction section.

The organization recommended here must often be adapted to meet the specific requirements of the report. For example, a report on a test of emissions from a smokestack may take the following form:

(1) Introduction
- test objectives
- brief process and control equipment description
- test dates and personnel
- observers

(2) Summary of results
- brief test method identification
- regulatory agency approval of method
- comments on process operation
- emission rate determined by the test
- emission rate limit given by law

(3) Process description
- describe process
- describe control equipment
- flow diagram of entire process
- charts and calculations of process production rates

(4) Testing methodology
- sampling scheme with drawing and dimensions of site and sample points
- description of sampling method
- description of analytical method
- modifications to methods and approved justification

(5) Results
- summary of data
- charts and tables
- example calculations

(6) Appendix

Note the similarity with the first listing, but the specific adaptation to the immediate requirements. Now returning to the first list, we will describe each element in more detail.

TITLE PAGE

The title page normally consists of: 1) the complete title, 2) the name of the person to whom the report is submitted, 3) the name of the writer (listed first) and his/her coworkers, 4) the date of submittal and 5) the organization issuing the report. This information is arranged so that the page has a balanced appearance. The title is always presented at the top of the page in capital letters. For the purposes of a lab report, item 5 can be the course number and name, department and University of Cincinnati. The form of the title page for a Master's Thesis is prescribed by the College of Engineering and an example title page is provided to students.

TABLE OF CONTENTS

List in the Table of Contents every title and major heading in the text and the number of the page on which it appears. Show the relationship of subdivisions through the use of indentations. For example, a lengthy discussion should be broken up into subsections, titled and shown as subdivisions indented under, e.g., "Discussion of Results," in the Table of Contents. List all tables, figures, sketches, graphs and photographs separately, either as they appear in the text or as separate appendices, at the end of the Table of Contents. If appendices are used to organize pictorial or tabular sheets, then the figure or table titles can be indented under the "Appendix" heading in the Table of Contents.

The first letter of each word should be capitalized, except for articles, prepositions and the like. Major report sections are preceded by Roman numerals, subsections by capital letters and any further subdivisions by Arabic numerals and lower case Roman numerals. The Abstract, Nomenclature, References and Appendices *are not* considered major report sections when organizing the Table of Contents. Tables are numbered with Roman numerals and figures with Arabic numerals.

ABSTRACT

The Abstract requires the most careful writing in the report. It is the *most* important part of the whole report. Its purpose is to immediately acquaint the reader with more of the content of the report than is contained in the title. It should be comprehensive, covering all phases of the investigation, but should not go into minor details in any particular phase. Qualitative statements or vague diction have no place in the Abstract; precise quantitative expression without unnecessary information is the aim. Specifically, an abstract must include the following:

- a statement introducing the subject matter to the reader

- what was done
- some selected results (the most important), using numerical values, if possible

Discussion of results is rarely included in the abstract. Usually, the essential information can be given in five or six well-chosen sentences. The abstract is usually the last part of the report to be written, even though it appears first in the completed report. Precautions:

- Do not refer to graphs or tables.
- Do not employ the exact phrasing of the text.
- Do not present poor results or those which you question.
- Do not cite to references.

INTRODUCTION

Some of the functions of the Introduction are to orient the reader, to state the problem, to delimit the field of investigation, to define special terms and to outline the method of attack.

In the Introduction, tell why the experiment was done, but state only its technical purpose; such remarks as "to acquaint the student" and "to give practice in laboratory technique" are not acceptable. Try to avoid trite phrases like, "The purpose of this experiment was. . . ."

THEORY

In your reports, it is seldom possible to review the theory and practice of a given problem exhaustively. The aim should be the development of the theoretical principles and a discussion, evaluation and criticism of the principle contributions in the literature ("literature review").

Elementary theory or equations available in standard undergraduate texts may be presented, referenced and numbered, but not derived. If an equation of the student's own derivation is used, its derivation is given in this section. Theoretical principles are stated in the present tense, while references to the experimental work of others are given in the past tense.

EQUIPMENT AND MATERIALS

This section is a written description of all pieces of equipment and materials used. References to sketches or tabulated dimensions are very helpful.

Sufficient quantitative details and measurements must be given, so that a technically trained reader could duplicate the experimental setup. For example, the nameplate data on the motors and control instruments, materials of construction, capacities, etc., are given. Careful description is made of any details that might influence the results of the experiment. First, the reader should be given a general description of the most important features of the equipment. Then, in subsequent paragraphs, the details are described. Avoid a description which merely proceeds from left to right; instead, relate the description to the functions of the equipment or to its key components. A list of equipment used is not a satisfactory description of the experimental setup!

EXPERIMENTAL PROCEDURE

Description of the procedure is analogous to that of the equipment. Again, the first paragraph is devoted to essential features of the entire operation. Subsequent paragraphs should develop these features so clearly and in such detail that a competent reader familiar with the subject matter could repeat the experiment. Specific equipment arrangements should be included where appropriate and a diagram used for illustration.

If possible, state the reason for using a particular procedure. For example, "As a precaution against atmospheric oxidation of the olefins, the system was flushed with nitrogen admitted through inlet, L."

Be sure to tell what was done and not what should have been done!

Significant observations should be included next to the particular step during which they occurred. Any interpretations are left to the discussion section.

RESULTS

Results are customarily gathered into the form of graphs, tables, or equations and grouped according to the particular variable studied. When no such grouping is possible, the results may consist of a list of concise statements summarizing the entire results of your work and details included in the appendix.

DISCUSSION OF RESULTS

Each graph, table, equation, or statement appearing in the Results section must be discussed thoroughly and concisely. In discussing the results, state the assumptions, important sources of error (quantitatively, when possible), the methods used in achieving the results and the extent of agreement with theory or previous work. Comparisons should be expressed quantitatively, e.g., as a percent deviation. Explain any exception and support your explanations by experimental data if possible. Tell why your graphs look the way they do. Your goal here is to analyze why something did or did not happen. You must try to be concise and avoid the tendency to write unnecessary information just to make the report bulky.

CONCLUSIONS

Conclusions are a series of (often numbered) sentences which answer the questions posed in stating the purpose of the experiment. They include only pertinent information and are based solely on data and results within the report. For example, the final prose section on friction in a duct might conclude: (1) "The measured values of friction in a duct presented in Table III were in error by no more than 18%." (2) "The observed values were consistently below those predicted." Be sure to distinguish between a fact and a conclusion. The former leads to the latter.

RECOMMENDATIONS

Recommendations are specific definite proposals for future work, e.g., suggested changes in equipment, study of new variables, possible experiments in related fields, a calculation procedure based on your experiments, or the best choice between several designs considered during your work. Like the Conclusions, the Recommendations are usually listed by number, and each consists of only a few sentences.

NOMENCLATURE

All symbols used in the text and theory should be arranged in alphabetical order followed by (in order) Greek symbols, subscripts and superscripts. For our purposes, symbols used in the appendices may be listed in the Nomenclature section, although this is generally not permissible in practice. Symbols should be standard, and physical quantities should never be represented by two different symbols in your report.

REFERENCES

A list of references should be presented according to a standard reference system. See, e.g., *Handbook for Authors of Papers in the Journals of the American Chemical Society* or, for legal references, *The Uniform System of Citation.* Be as specific as possible, including page number of an equation or graph in the referenced work, chapter number if ideas used in the reference are numerous, or just the book reference if referenced material was scattered throughout the book.

APPENDICES

Experimental data, graphs and figures, sample calculations, tabulated calculation results, calibration curves, the code of any computer program used, manufacturer's bulletins and any other information or material not included in the report are presented in the appendix.

Experimental and calculated data are best presented in tabular form, taking care to clearly distinguish between experimental and calculated data. Tables, figures and graphs should be clearly labeled with the variable being measured and the units of every quantity should be stated. The original data sheets, photocopied from your notebooks, are always inserted into the appendix of a lab report. They are generally not presented in a Master's Thesis/Project Report.

The purpose of the sample calculation is to show exactly how observed data were transformed into calculated data, and finally into the results. Calculations need to be given for only one complete experiment in a series of experiments. Actual data must include complete specification of the units. A symbolic (formula) presentation of the calculations generally precedes the actual production of numbers. Be sure that there is enough explanation, so the reader can follow every number in every step of the calculation. If graphical techniques are used, a full explanation of *each* step is necessary. State which particular experiment was singled out for demonstration. The sources for any physical data should be given.

STANDARD FORMS

EQUATIONS

(1) In printed reports, it may be necessary to leave spaces so that equations can be inserted in ink.

(2) Short equations can sometimes be run with the text. Example: can be predicted by $V^2 = 2hg$ where V is the. . . .

(3) In general, centering of the equations on the page is preferred. Explanations of the symbols are tabulated below it (unless the symbols are given in a separate table of nomenclature). Example:

$$ha + \frac{U^3a}{2g} + \frac{Pa}{Ca}W - F = \frac{U^2b}{2g}\frac{Pb}{Ca} \qquad (12)$$

where:

h = height of feed at points a and b, ft
w = work added by a pump, ft-lb
etc.

(4) Words like *therefore, but* and *substituting* between or preceding equations should be placed flush with the left-hand margin.

(5) When equations are numbered, the numerals appear flush with the right-hand margin. The numerals are placed in parentheses or brackets.

$$V^2 = 2gh \qquad (1)$$

(6) In referring to equations, it is advisable to omit the parentheses or brackets. Correct: "The equation $V^2 = 2gh$ is derived from Equations 1 and 2"; *not* "Equations (1) and (2)."

REFERENCES TO LITERATURE

(1) Put reference numbers or publication dates referring to the bibliography in parentheses immediately following the author's name.

(2) When referring to authors in the body of the report give only the last name. *Exception:* Famous scientists such as J. J. Thompson, J. Willard Gibbs, Lord Rayleigh, etc.

(3) If there are more than two authors, state only the name of the first author and add et al., as, for example, "Jones et al. (1981), report that the effect. . . ." Note that the subject is plural.

(4) Use a consistent standard form for your references, such as the *Chicago Manual of Style*, *Turabian* or the *ACS Handbook for Authors*.

ABBREVIATIONS

(1) The general rule is to avoid the use of abbreviations in the writing itself and to use too few, rather than too many abbreviations.

(2) Units of measure are abbreviated only when preceded by numerals as, 18 ft, 3 in, 700 rpm, 10 kW or 15%.

(3) Abbreviations are generally given in the singular as 3 lb, not 5 lbs; 3 in, not 3 ins.

(4) Abbreviations are given in lowercase letters. *Exceptions:* BTU (but Btu is acceptable), US gal, Eq. 17, Vol. 4, No. 9.

(5) Such signs as ′ for feet, ″ for inches, @ for at and / for per are *not* used in the text, but are used in tables and computations when space must be saved.

(6) Periods are used after abbreviations of units *only* at the end of a sentence. The following are correct:
- three 3-ft pipes
- a line 3 ft long
- 20-hp motor
- measured in horsepower

- at 500 rpm
- 64% recovery
- 18 kW supply
- every 5 minutes

(7) Data is a plural word.

NUMBERS

(1) The general rule is to write in numerals all numbers above nine.
(2) Numerals are used with measures of quantity and with such terms Fig., Report and Table, as shown in the rules for abbreviations.
(3) Decimals and complex fractions are given in numerals as 2½, 1.25 and 0.0109.
(4) When many numbers are given in the same section of writing, numbers are used for all.
(5) Numbers used as approximations or as first words of sentences are given in words: ". . . about twenty feet. "One hundred four tests were made. Better, rewrite the sentence."
(6) Never start a sentence with numerals.
(7) Be careful to report only *significant figures*.
(8) Express approximations as fractions, rather than as percent. The latter implies precision. This is correct: "When the tank was half full." Incorrect: "When the tank was 50 percent full."

HYPHENS

(1) Two adjacent nouns are hyphenated if they express a single conception and if, without the hyphen, the singleness of their meaning were not immediately clear as feed-water (or feedwater), tool-steel, feed-pipe, safety-valve. (This rule is given various interpretations.)
(2) Two words compounded to form an adjective are hyphenated, for example, steam-jacketed kettle.
(3) Hyphens are used with words beginning with non, semi, pseudo, etc., and with measures as 3-in valve, 15-min test, when it forms an adjective.

TABLES, FIGURES AND GRAPHS

General: Whether these are placed in the body of the report or in the appendix is largely a matter of individual judgment. If located in the main text, they should follow the page on which they are first referred to. The pages on which they appear are numbered in sequence with the other pages of the report.

In addition to an informative title, sufficient subtitles or legends should be provided, so that each figure or graph is self-explanatory. There should be no need for the reader to refer back to the text for necessary information. This is very important.

Never have any writing that presents itself upside down when the report is held in the normal reading position, or when it is held with the bound edge at the top.

Figures and tables requiring a page size larger than 8½ × 11 inches should be bound to such a page by a heavy binding strip. Care should be taken to fold the large sheet, so as to prevent wear on folded edge and the corners.

Figures: It is usually desirable to use separate sheets of paper for figures. The figure number and the title should be typed at the bottom of the page, as for example: Fig. 10. Reaction Chamber for Oxidation of Methane.

In complicated drawings with many parts, the components should be identified by numbers inside circles with arrows leading to the part concerned. The text should refer to these parts by name and by number, for example, ". . . The catalyst support (Item 16) was constructed. . . ." It is necessary to use "Item" to distinguish these references from bibliographical references. Figures are numbered serially with Arabic numerals.

Tables: If tables are less than a page long, they may be run in with the text. Avoid, as far as possible, continuing tables from one page to another. Abbreviations may be used in the column and row headings, but the nature of the quantities being listed must be clear. Tables are numbered serially with Roman numerals.

Graphs: Graphs should occupy the central portion of a separate page with plenty of margin all around. The axes must be clearly labeled, and the scales should be placed along the respective axes.

Unless there is good reason to omit it, the origin should appear on all graphs. The zero lines should be sharply distinguished from the other coordinate lines or when logarithmic coordinates are used, the limiting lines of the diagram should each be some power of ten on the logarithmic scale.

At each experimental point, make a clear locational symbol, a circle, triangle, cross, etc. Pass a smooth curve through the points, drawing the most probable line. Usually the most probable line is a simple, smooth curve with "saw teeth." The line may not pass through any of the observed points if the scatter of data is appreciable. In general, do not have more than one or two points of inflection on your curves. Where extrapolation or interpolation is doubtful, draw the curve as a dotted line.

The use of smooth curves becomes illogical for certain observations, and in such cases, a straight edge is used to connect the actual data points. The resulting plot may have a sawtooth appearance. For example, if one were plotting price of sulfuric acid vs. day of the year, a jagged line (not a smooth one) would result. A few classes of calibration curves fall in this category.

Broken lines or colored lines may also be used. The individual lines must be clearly labeled or else an informative "key" must be shown somewhere on the same sheet.

No data should appear on a graph which are not tabulated elsewhere in the report. Graphs are numbered serially with Arabic numerals.

TEMPERATURE

THE FAHRENHEIT AND CELSIUS SCALES

The range of units on the Fahrenheit scale between the freezing and boiling point of water at one atmosphere (atm) pressure is 180 (212°F−32°F = 180°F); on the Celsius scale, the range is 100 (100°C−0°C = 100°C). Therefore, each Celsius degree is equal to 9/5 or 1.8 Fahrenheit degrees. To be able to convert from one system to the other, the following equations can be used:

$$°F = 1.8(°C) + 32 \tag{1.1}$$

$$°C = \frac{(°F - 32)}{1.8} \tag{1.2}$$

where:

°F = degrees Fahrenheit
°C = degrees Celsius

ABSOLUTE TEMPE⁻ATURE

Gas volume can be determined as a function of temperature at a constant pressure for a given mass of material. If gas fraction plots of volume vs. temperature lines are extrapolated to a volume of zero, they all intersect at a common temperature (−273.15°C or −459.67°F). This is the temperature at which a gas, if it did not condense, would theoretically have a volume of zero. This temperature (−273.15°C or −459.67°F) is called *absolute zero.* Another temperature scale, developed by and named after English physicist Lord Kelvin, begins at absolute zero and has temperature intervals equal to Celsius units. This absolute temperature scale is in units of *degrees Kelvin* (K). A similar scale was developed to parallel the Fahrenheit scale and is called the *Rankine scale* (°R). The following formulas can be used to convert temperatures to their respective absolute scales.

$$°K = °C + 273.16$$

$$°R = °F + 459.67$$

PRESSURE

DEFINITION OF PRESSURE

A body may be subject to three kinds of stress: shear, compression and tension. Fluids are unable to withstand tensile stress; hence, they are subject to shear and compression only. *Unit compressive stress in a fluid is termed pressure* and is expressed as *force per unit area*. In metric units, this is Newton/m^2; and in English units, it is lb/in.^{2} (psi). Pressure is equal in all directions at a point within a volume of fluid and acts perpendicular to a surface.

BAROMETRIC PRESSURE

Barometric pressure and atmospheric pressure are synonymous. These pressures are measured with a barometer and are usually expressed as inches or millimeters of mercury. *Standard barometric pressure is the average atmospheric pressure at sea level, 45° north latitude at 35°F and is equivalent to a pressure of 14.696 lbs-force per square inch exerted at the base of a column of mercury 29.921 inches high (in the English System). In the metric system, standard barometric pressure is equivalent to a pressure of 1033.23 grams-force per square centimeter exerted at the base of a column of mercury 760 mm high*. Weather and altitude are responsible for barometric pressure variations.

TORRICELLI BAROMETER

The Torricelli, or mercurial, barometer was first used by one of Galileo's students, Torricelli, in 1643. A mercurial barometer is made by sealing a tube, about 32 inches long, at one end. The tube is filled with mercury. The mercury in the tube will fall until the weight of the mercury in the tube is equal to the force of the air pressure on the mercury in the container. As shown in Figure 1.1, the manometer and the mercurial barometer work on the same principle—atmospheric pressure being measured with reference to a vacuum.

FORTIN BAROMETER

Since the mercurial barometer is the most accurate measurement (calibration uncertainty of 0.001 to 0.03% of reading) of atmospheric pressure, it is still in

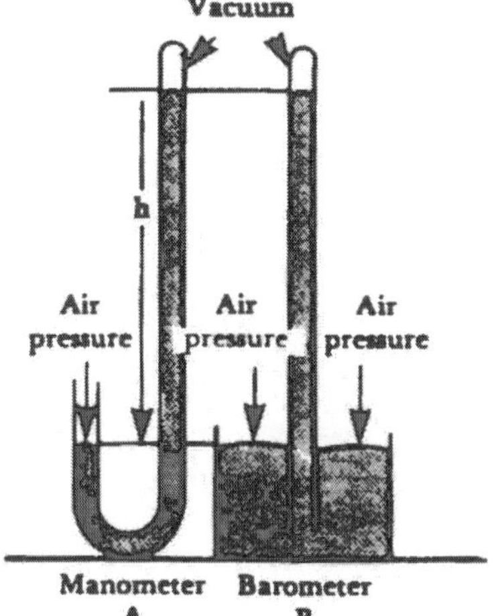

Figure 1.1 The manometer and mercurial barometer.

wide use today. The most common modified version of the mercurial barometer is the Fortin type shown in Figure 1.2.

The height of the mercury column in a Fortin barometer is measured from the tip of the ivory index point (see Figure 1.3) to the top of the mercury column. The mercury level in the glass cylinder (ambient-vented cistern) is adjusted until the ivory index point just pricks the surface of the mercury. This is done by turning the datum-adjusting screw. Then the vernier scale is adjusted until the bottom of it is even with the top of the mercury meniscus. After the vernier scale is adjusted, the height of the mercury column is read.

A typical vernier scale is shown in Figure 1.4. The barometric pressure indicated in the figure is determined in the following way.

The bottom of the vernier scale indicates not only the integer component of the barometric pressure, but also the tenths components–in this case, 29.9. The hundredths component is indicated by the match between the outer scale and the vernier–in this case, 0.04. The readings are totaled to determine the barometric pressure: 29.9 + 0.04 = 29.94 in. Hg. The equivalent metric reading is 76.05 cm.

ANEROID BAROMETER

The aneroid barometer is usually not as accurate as a Torricelli barometer. However, aneroid barometers are more widely used, because they are smaller, more portable, less expensive and easier to adapt to recording instrumentation than are Torricelli barometers.

The aneroid barometer usually consists of a metal chamber, bellows, or sylphon (accordion-like) cell that is partially evacuated. A spring is used to

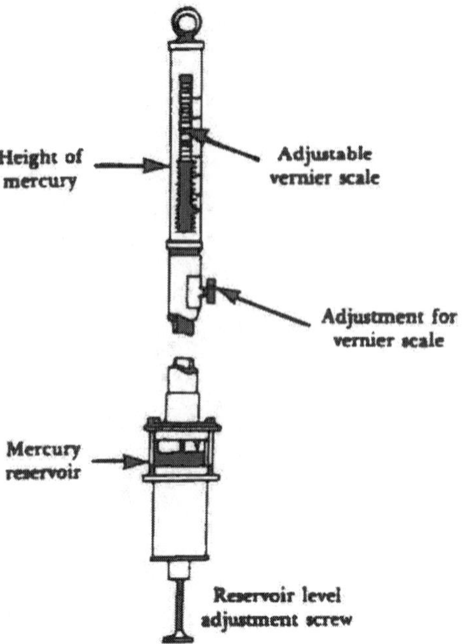

Figure 1.2 Fortin barometer.

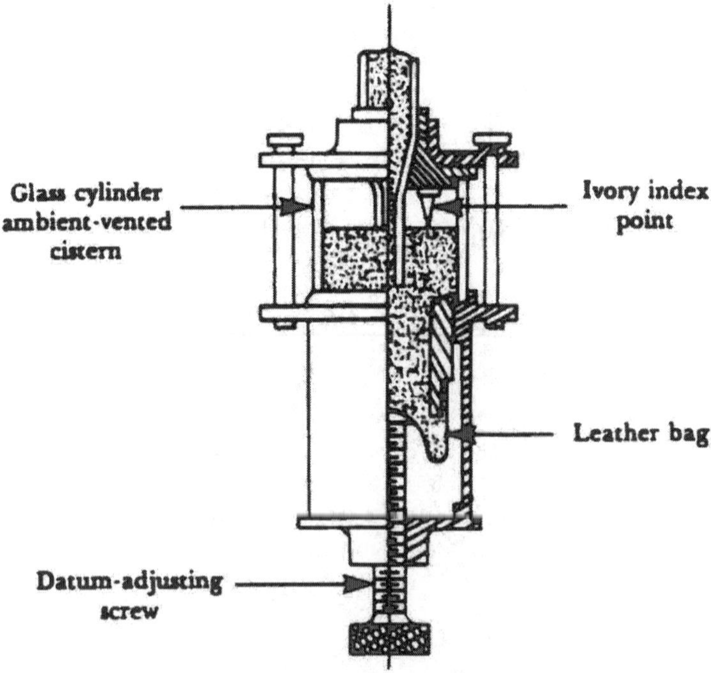

Figure 1.3 Blow-up of Fortin barometer.

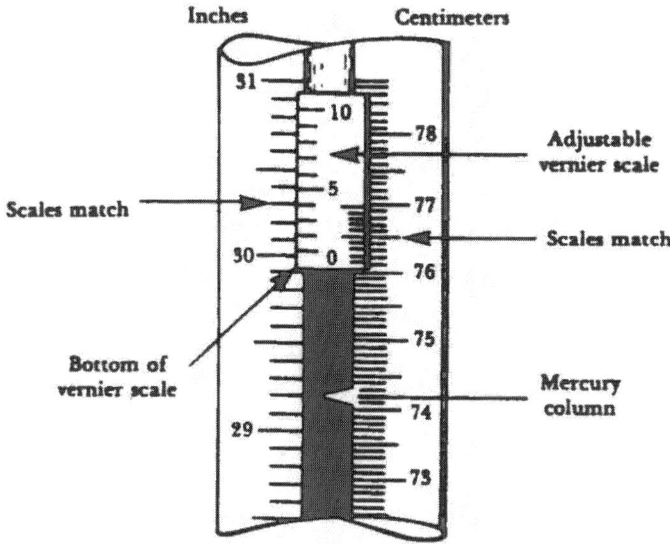

Figure 1.4 Blow-up of vernier scale.

keep the metal chamber from collapsing (see Figure 1.5). The width of the chamber is determined by the balance between the spring and the force exerted by the atmosphere. The width of the chamber is indicated by a pointer and scale that can be calibrated to read directly in units of pressure (i.e., millimeters or inches of mercury, etc.). The pointer movement can be amplified by using levers. Readout systems can vary from visual scales to recording devices. The combination of an aneroid barometer and an automatic recording device is called a barograph.

GAUGE PRESSURE

Gauges indicate the pressure of the system of which they are a part relative to ambient barometric pressure. If the pressure of the system is greater than the pressure prevailing in the atmosphere, the gauge pressure is expressed as a positive value; if smaller, the gauge pressure is expressed as a negative. The term "vacuum" designates a negative gauge pressure.

The abbreviation "g" is used to specify a gauge pressure. For example, *psig* means pounds-force per square inch gauge pressure.

ABSOLUTE PRESSURE

Because gauge pressure (which may be either positive or negative) is the pressure relative to the prevailing atmospheric pressure, the gauge pressure, added algebraically to the prevailing atmospheric pressure (which is always

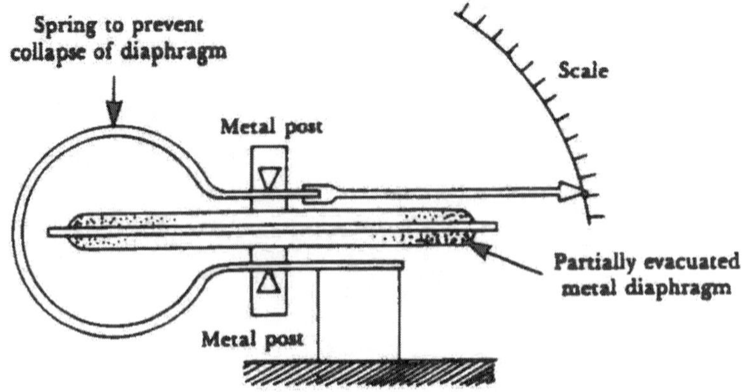

Figure 1.5 Aneroid barometer.

positive), provides a value that is called "absolute pressure." The mathematical expression is:

$$P = P_b + p_g \quad (1.3)$$

where:

P = absolute pressure
P_b = atmospheric pressure
p_g = gauge pressure

Note: P, P_b, and p_g must be in the same units of pressure before they can be added (i.e., all must be in inches of mercury, mm of mercury, etc.).

The abbreviation *"a"* is sometimes used to indicate that the pressure is absolute. For example, *psia* means pounds per square inch absolute pressure. Equation 1.3 allows conversion of one pressure system to the other.

FLOW METERS

The most popular devices for measuring flow rate are the rate meters. Rate meters measure, indirectly, the time rate of the fluid flow through them. Their response depends on some property of the fluid related to the time rate of the flow.

FIXED AREA METERS

Variable Pressure Meters—Head Meters

Head meters are those in which the stream of fluid creates a significant pressure difference that can be measured and correlated with the time rate of flow. The pressure difference is produced by a constriction in the stream of flow causing a local increase in velocity.

Orifice Meter—Noncritical—Secondary Standard

An orifice meter can consist of a thin plate having one circular hole coaxial with the pipe into which it is inserted (Figure 1.6). Two pressure taps, one upstream and one downstream of the orifice, serve as a means of measuring the pressure drop, which can be correlated to the time rate of flow. Watch jewels, small bore tubing and specially manufactured plates or tubes with small holes have been used as orifice meters. The pressure drop across the orifice can be measured with a manometer, magnehelic, or pressure gauge.

Flow rates for an orifice meter can be calculated using Poiseuille's Law; however, this is not done for practical use. Instead the orifice meter is usually calibrated with either a set or dry test meter, or a soap-bubble meter.

Calibration curves for orifice meters are nonlinear in the upper and lower flow rate regions and are usually linear in the middle flow rate region.

Orifice meters can be made by laboratories with a minimum of equipment. They are used in many sampling trains to control the flow. Care must be exercised to avoid plugging the orifice with particles. A filter placed upstream of the orifice can eliminate this problem. Orifice meters have long been used to measure and control flows from a few ml/min to 50 l/min.

Orifice Meter—Secondary Standard

If the pressure drop across the orifice (Figure 1.6) is increased until the downstream pressure is equal to approximately 0.53 times the upstream pressure (for air and some other gases), the velocity of the gas in the constriction will become acoustic, or sonic. Orifices used in this manner are called critical orifices. The constant 0.53 is purely a theoretical value and may vary. Any further decrease in the downstream or increase in the upstream pressure will not affect the rate of flow. As long as the 0.53 pressure relationship exists, the flow rate remains constant for a given

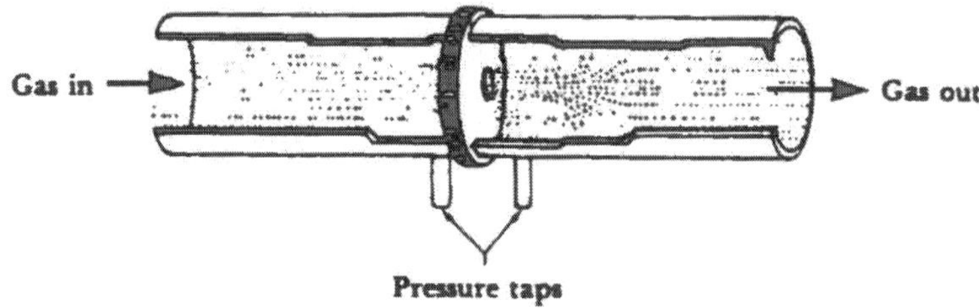

Figure 1.6 Orifice meter.

upstream pressure and temperature, regardless of the value of the pressure drop. The probable error of an orifice meter is in the neighborhood of 2%.

Only one calibration point is needed for a critical orifice. The critical flow is usually measured with a soap-bubble meter, or a wet or dry test meter. Corrections for temperature and pressure differences in calibration and use are made with the following formula:

$$Q_2 = Q_1 \left[\frac{P_1 T_2}{P_2 T_1} \right] \tag{1.4}$$

where:

Q = flow
P = pressure
T = temperature in K
1 = initial conditions
2 = final conditions

The same formula can be used to correct orifice meter flows to standard conditions by substituting $P_2 = 760$ mm Hg and $T_2 = 298$ k. Note the square root function of T and P. *Any time that rate meters are corrected for T and P, this square root function is needed.*

Critical orifices are used in the same types of situations as noncritical orifices. Care must also be taken not to plug the orifice.

Venturi Meter—Secondary Standard

The venturi meter consists of a short cylindrical inlet, an entrance cone, a short cylindrical throat and finally a diffuser cone (Figure 1.7). Two pressure taps, one in the cylindrical inlet and one in the throat, serve to measure the pressure drop. There is no abrupt change of cross section as with an orifice; thus the flow is guided both upstream and downstream, eliminating turbulence and reducing energy losses. Venturi meters are, of course, more difficult to fabricate. The probable error of a venturi is 1%.

The venturi meter is calibrated in the same manner as the orifice meter. The calibration curve generated plots pressure drop across the venturi versus flow rates determined by the standard meter.

VARIABLE AREA METERS

The variable area meter differs from the fixed orifice; the pressure drop across it remains constant while the cross-sectional area of the constriction

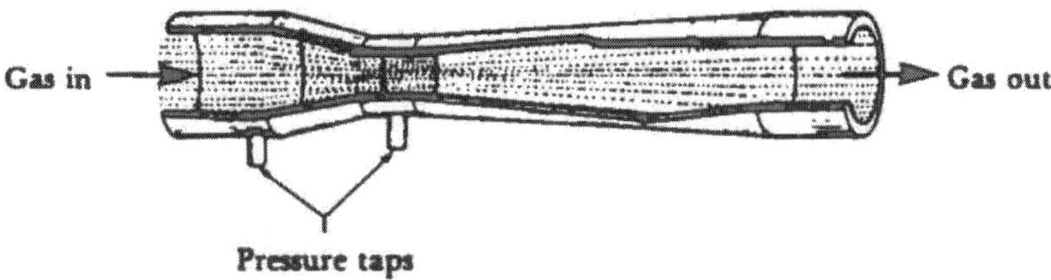

Figure 1.7 Venturi meter.

(annulus) changes with the rate of flow. A rotameter is an example of a variable area meter.

Most rotameters are used and calibrated at room temperature with the downstream side at atmospheric pressure. Corrections for pressure and temperature variations can be made using the previously mentioned formula:

$$Q_2 = Q_1 \left[\frac{P_1 \times T_2}{P_2 \times T_1} \right]^{1/2} \tag{1.5}$$

If a gas is measured with a different density from the calibration gas, the flow rate can be corrected using the following formula:

$$Q_1 = Q_2 \left[\frac{P_2}{P_1} \right] \tag{1.6}$$

where:

Q_1 = flow rate with gas 1
Q_2 = flow rate with gas 2
P_2 = density of gas 2
P_1 = density of gas 1

Because corrections of this type are cumbersome and add inaccuracies, rotameters are usually calibrated under normal operating conditions against a primary or intermediate standard.

Rotameters are the most widely used laboratory and field method for measuring gas or liquid flow. Their ease of use makes them excellent for spot flow checks. Many atmospheric sampling instruments use rotameters to indicate the sample flow rate. With proper calibration, the rotameter's probable error is 2 to 5%.

VELOCITY METERS

Velocity meters measure the linear velocity or some property that is proportional to the velocity of a gas. Several instruments exist for measuring the velocity of a gas; we will discuss only the pitot tube and the mass flow meter. Volumetric flow information can be obtained from velocity data, if the cross-sectional area of the duct is known, using the following formula:

$$Q = Av \tag{1.7}$$

where:

Q = volumetric flow rate (m^3/min)
A = cross-sectional area (m^2)
v = average velocity (m/min)

Pitot Tube

The pitot tube is a simple pressure-sensing device used to measure the velocity of a fluid flowing in an open channel. The complexity of the underlying fluid-flow principles involved in a pitot tube gas-velocity measurement is not apparent in the simple operation of this device. The pitot tube should, however, be considered and treated as a sophisticated instrument.

The pitot tube actually measures the velocity pressure (Δp) of a gas stream. Gas streamlines approaching a round object placed in a duct flow around the object except at point "P_+," where the gas stagnates and the stagnation pressure (P_+) is found (Figure 1.8 and Figure 1.9). P_+ is the sum of kinetic pressure plus static pressure. The static pressure in a gas stream is defined as the pressure that would be indicated by a pressure gauge if it were moving along with the stream, or "static," with respect to the fluid.

The difference between the stagnation pressure (P_+) and the static pressure (P_s) is the velocity pressure differential (Δp). This is shown in Figure 1.9.

Bernoulli's Theorem relates pitot tube velocity pressure (Δp) to gas velocity in the following equation:

$$v = K_p C_p \left[\frac{T\Delta p}{PM}\right]^{1/2} \tag{1.8}$$

where:

v = velocity of the gas stream, ft/sec
T = absolute temperature, R(°F + 460)
P = absolute pressure, in Hg
M = molecular weight of the gas, lb/lb-mole
Δp = velocity pressure, in H_2O
K_p = 85.49 ft/sec [(lb/lb·mole)(in Hg)/(in H_2O)(°R)]$^{1/2}$
C_p = pitot tube coefficient, dimensionless

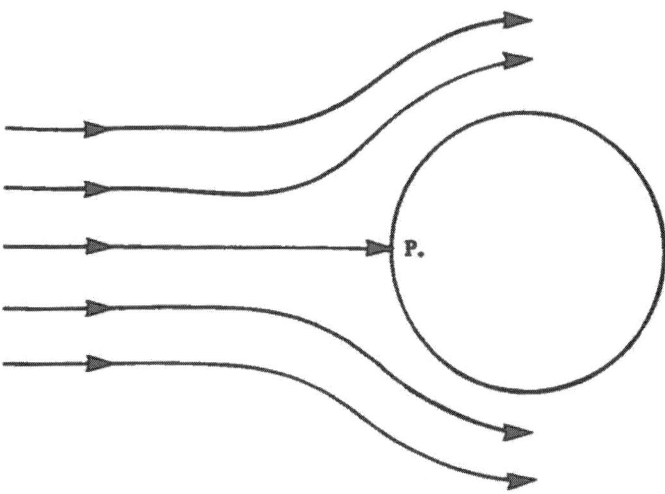

Figure 1.8 Gas stagnation against an object.

Pitot tubes are used extensively in ventilation work to measure air flow in ducts. Literature sources describe pitot tubes in detail. The standard and S-type pitot tubes are the most commonly used.

Standard Pitot Tube—Primary Velocity Standard

The standard pitot tube (Figure 1.10) consists of two concentric tubes. The center tube measures the stagnation or impact pressure, and the static pressure is measured by the holes located on the side of the outer tube. The pitot tube must be placed in the flowing air stream so that it is parallel with

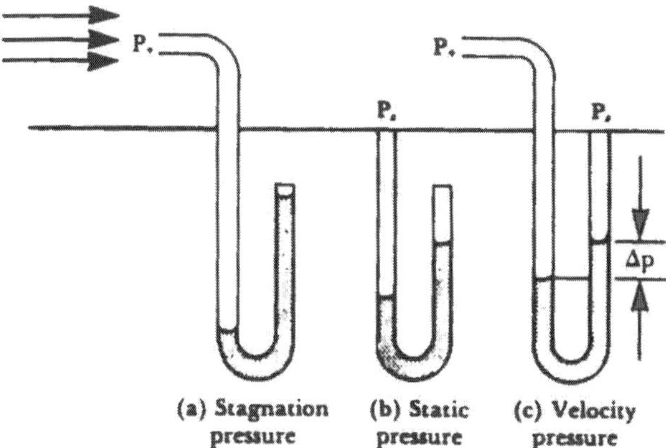

Figure 1.9 Pitot tube pressure components.

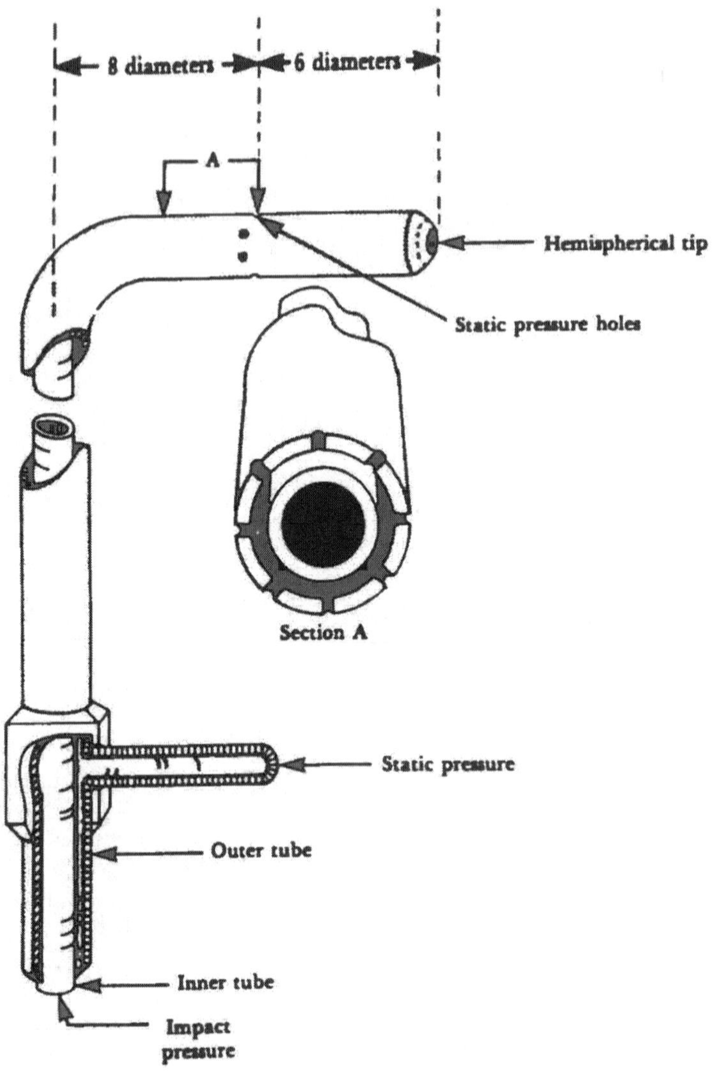

Figure 1.10 Standard pitot tube.

the streamlines. The velocity pressure differential (ΔP) can be measured with a U-tube manometer, an inclined manometer, or any suitable pressure-sensing device. Only velocities greater than 2500 ft/min can be measured with a U-tube manometer, but flows as low as 600 ft/min can be measured with a carefully adjusted inclined manometer. Standard pitot tube velocity pressures are typically 0.14 inches of water at 1500 ft/min and 0.56 inches of water at 3000 ft/min.

The standard pitot tube was first calibrated against an orifice meter using Bernoulli's Theorem. Repeated calibrations proved that different standard pitot tubes have the same characteristic flow calibration. If the static pressure holes are six outer tube diameters from the hemispherical tip and eight outer tube diameters from the bend (Figure 1.10), then the C_p value in Equation 1.8 is very close to 1.0 (it actually is 0.99).

Standard pitot tubes can be used to measure linear velocity in almost any situation except in particle-laden gas streams. The particulate matter will foul the carefully machined tip and orifices. The velocity of gas streams with high particulate matter concentrations can be measured better with an S-type pitot tube.

S-Type Pitot Tube

The S-type pitot tube consists of two identical tubes mounted back to back (Figure 1.11). The sampling ends of the tubes are oval with the openings parallel to each other. In use, one oval opening should point directly upstream, the other directly downstream. The tubes should be made of stainless steel or quartz if they are used in high temperature gas streams. The alignments shown in Figure 1.11 should be checked before use or calibration, as this may cause variations in the calibration coefficient (C_p).

Calibration of the S-type pitot tube is performed by comparing it to a standard pitot tube. Both the standard and S-type pitot tubes are placed alternately into a constant air flow. Pressure readings are taken for the standard pitot tube and for leg A of the S-type tube, facing the direction of flow and leg B, facing the direction of flow (Figure 1.11). The pitot tube coefficient (C_p) is calculated using the following formula:

$$C_{p(s)} = C_{p(\mathrm{STD})}\left(\frac{\Delta P_{\mathrm{STD}}}{\Delta P_S}\right)^{1/2} \tag{1.9}$$

The average C_p is calculated from several readings and should have a

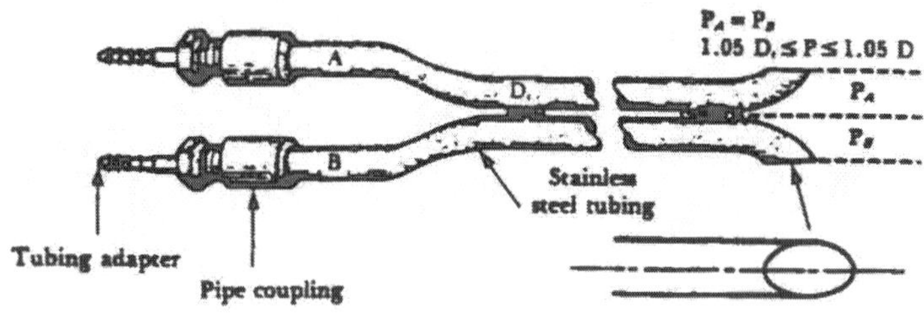

Figure 1.11 S-type pitot tube.

value of approximately 0.84. The C_p for leg A and leg B should differ by less than 0.01. The C_p value can then be used to calculate velocity by Equation 1.8. Note that when properly constructed, a C_p of 0.840 may be assumed without calibration.

The S-type pitot tube maintains calibration in abusive environments. The large sensing orifices minimize the chance of plugging with particulates. The S-type pitot tube also gives a high manometer reading for a given gas velocity pressure, which can be helpful at low gas velocities. These features make the S-type pitot tube the most frequently used instrument to measure stack gas velocity.

MASS FLOW METERS—SECONDARY VELOCITY STANDARD

Mass flow meters work on the principle that when a gas passes over a heated surface, heat is transferred from this surface to the gas. The amount of current required to keep the surface at a constant temperature is a measure of the velocity of the gas. Since the amount of heat transferred depends on the mass and velocity of the gas, these meters measure mass flow rate.

Atmospheric sampling applications of the mass flow meter are usually limited to the measurement of volumetric flow. Since these devices measure mass flow directly, they should be calibrated against a primary, intermediate, or secondary volumetric standard. The standard meter flow is corrected to standard conditions and compared to the mass flow rate measured. No corrections for temperature and pressure need to be made to the mass flow meter readings. Calibration must be done with the same gas as will be measured in use, because different gases have different thermal properties.

Mass flow meters are most often used for flow measurement or as calibration transfer devices in the field and laboratory. Their insensitivity to temperature and pressure makes them a useful tool for standard conditions measurement.

VOLUME FLOW METERS

Wet Test Meter—Intermediate Standard

The wet test meter consists of a series of inverted buckets or traps mounted radially around a shaft and partially immersed in water (Figure 1.12). The location of the entry and exit gas ports is such that the entering gas fills a bucket, displacing the water and causing the shaft to rotate due to the lifting action of the bucket full of air. The entrapped air is released at the upper portion of the rotation and the bucket again fills with water. In turning, the drum rotates index pointers that register the volume of gas passed through the meter.

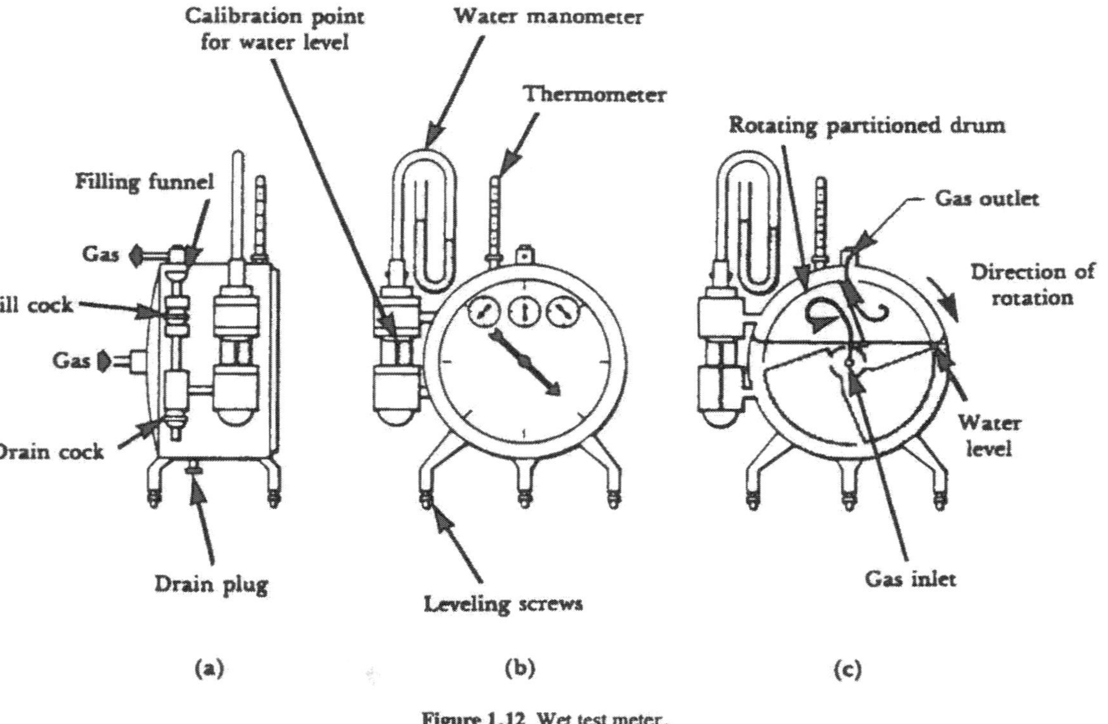

Figure 1.12 Wet test meter.

After the meter is leveled, the proper water level is achieved by using the filling funnel, fill cock and drain cock to bring the meniscus of the water in touch with the tip of the calibration index point. The calibration gas should be passed through the meter for one hour to saturate the water with the gas. The water in the meter should be at the same temperature as the surrounding atmosphere. If any water is added, sufficient time must be allowed for complete equilibration.

Wet test meters are used as transfer standards because of their high accuracy (less than ±1%). Because of their bulk, weight and equilibration requirements they are seldom used outside a laboratory setting. Wet test meters are useful for laboratories that need an accurate standard yet do not have the funds or space for a spirometer or mercury-sealed piston. Wet test meters can be used to measure flow rates up to 3 rev/min, at which point the meter begins to act as a limiting orifice and obstructs the flow. Typical ranges of wet test meters are 1, 3 and 10 l/rev.

Dry Test Meter—Intermediate Standard

Dry *test* meters are an improvement over the more common dry *gas* meters (Figure 1.13). Dry *gas* meters (a secondary standard) are most commonly used in residential and industrial settings to measure gas flow (e.g., natural gas). The dry *test* meter (an intermediate standard) works on the same principle as the dry *gas* meter (a secondary standard), but a different indexing method (readout) makes it more accurate (usually ±1 to 2% when new). The dry *test* meter shown in Figure 1.14 shows the new readout mechanism.

The interior of the dry test meter contains two or more movable partitions, or diaphragms, attached to the case by a flexible material so that each partition may have a reciprocating motion (Figure 1.15). The gas flow alternately inflates and deflates each bellows chamber, simultaneously actuating a set of slide valves that shunt the incoming flow at the end of each stroke. The inflation of the successive chambers also actuates, through a crank, a set of dials that register the volume of gas passed through the meter.

The dry test meter is calibrated against a spirometer, mercury-sealed piston, or displacement bottle similar to the wet test meter. One big advantage of the dry test meter over the wet test meter is that no correction for water vapor is needed. If the dry test meter is off calibration by more than 2%, it can be corrected by adjustment of the meter linkage. If linkage adjustment cannot correct the problem, then the dry test meter must be returned to the manufacturer for repairs.

Dry test meters are used in the field as well as in laboratory calibrations. Since the dry test meter does not contain water, it is lighter and easier to use

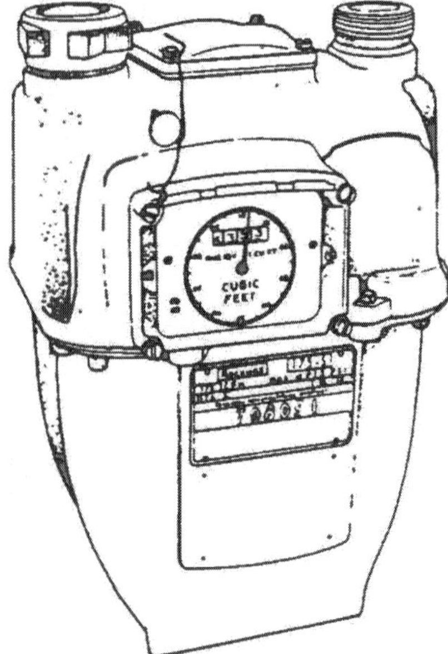

Figure 1.13 Dry gas meter.

than the wet test meter. Alsc he dry test meter is more rugged than the wet test meter. Accuracy of the dry test meter does, however, worsen with age.

GASEOUS SAMPLING

Gases can be sampled directly to an instrument or they may be sampled by using a holding bag and analyzed later. They also may be absorbed or reacted for indirect analyses. Several of these procedures are noted here.

GAS BAG INTEGRATED SAMPLES

Gas bags and tubing may be used to obtain a sample of gases. The bags and tubing must not react with the gas and chemicals must not evolve from the bags or tubing into the gas sample. The bags may be plastic bags with at least 1/4 to 1/2 ft^3 capacity made of aluminized Mylar, Tedlar, or equivalent. Also required may be a leakless vacuum pump or squeeze bulb, to transfer sample to the bag.

If a plastic bag is used, set up the equipment and check for leaks. Connect the pump, or the squeeze bulb, and draw the gas into the bag. In some cases, it may be desirable to draw an integrated sample. This is done by filling the bag in increments spread equally with time over the sampling period.

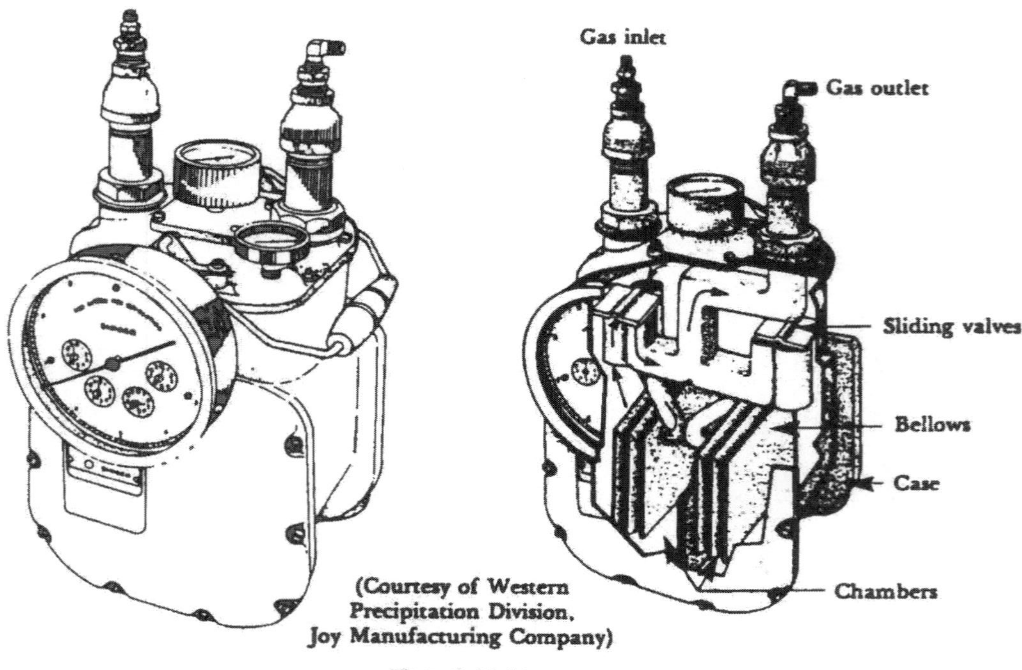

(Courtesy of Western Precipitation Division, Joy Manufacturing Company)

Figure 1.14 Dry test meter.

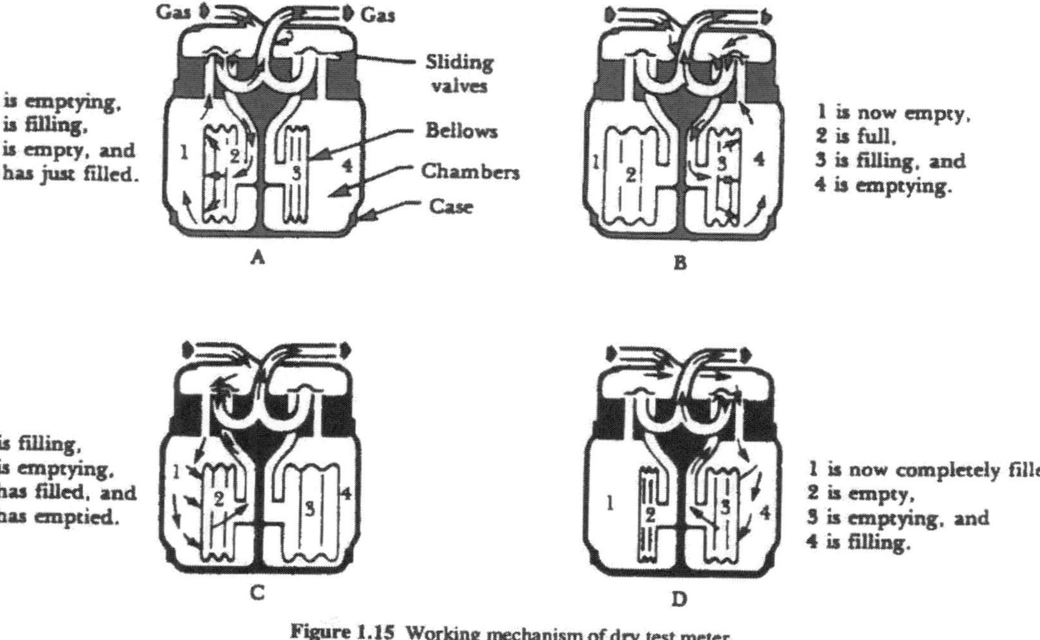
1 is emptying,
2 is filling,
3 is empty, and
4 has just filled.
Gas
Gas
Sliding valves
Bellows
Chambers
Case
1
2
3
4
A
1 is now empty,
2 is full,
3 is filling, and
4 is emptying.
1
2
3
4
B
1 is filling,
2 is emptying,
3 has filled, and
4 has emptied.
1
2
3
4
C
1 is now completely filled
2 is empty,
3 is emptying, and
4 is filling.
1
2
3
4
D

Figure 1.15 Working mechanism of dry test meter.

ABSORPTION OF GASES

Absorption of pollutants in various media plays an important role in air pollution monitoring. It is particularly important in the wet-chemical methods of analysis. Before the advent of continuous monitoring instrumentation, techniques employing absorption were the most inexpensive and up-to-date methods available.

Absorption is the process "of transferring one or more gaseous components into a liquid or solid medium in which they dissolve." Absorption of gaseous pollutants in solution is frequently utilized in atmospheric sampling because of the numerous methods available to analyze the resulting solution. These methods include photometric, conductimetric and titrimetric techniques. Details of sampling and analysis of specific gaseous pollutants by absorption are given elsewhere. This discussion concentrates on a description of the gas-liquid absorption process and factors affecting collection efficiency. Devices frequently utilized in gas-liquid absorption and several current applications are also discussed.

Types of Absorption

In gas-liquid absorption the collecting liquid (i.e., the absorbent) may change either chemically or physically, or both, during the absorption process. In gas-liquid absorption sampling, two types of absorption have been recognized: (1) physical absorption and (2) chemical absorption.

A typical chemical absorption process would involve drawing a volume of air through a solution that reacts with the gaseous contaminant to form a nongaseous compound; for example, an acid mist is drawn through a volume of sodium hydroxide. The acid reacts with the base to form a stable salt. Titration of the unreacted base with standard acid indicates the quantity of pollutant reacted.

Physical Absorption

Physical absorption involves the physical dissolving of the pollutant in a liquid. The process is usually reversible in that the pollutant exhibits a relatively appreciable vapor pressure. The solubility of the pollutant in a given absorbent is dependent on the partial pressure of the pollutant in the atmosphere and the temperature and purity of the absorbent. An ideal solvent would be relatively nonvolatile, inexpensive, noncorrosive, stable, nonviscous, nonflammable and nontoxic. In many cases, distilled water fulfills many of these characteristics and is used as the solvent for collecting some gases. The suitability of distilled water for several selected gases is presented in Table 1.1.

The physical absorption process involves collecting the pollutant by solution in the absorbent. The solution is then analyzed for pollutant concentration by a convenient analytical method. In general, low efficiency will be obtained for physical absorption unless the pollutant is very soluble and the ratio of dissolved gas to liquid volume is small. For this reason, physical absorption is rarely the only absorption process involved in collecting gaseous pollutants.

Chemical Absorption

In contrast to physical absorption, chemical absorption is a process that involves a liquid absorbent that reacts with the pollutant to yield a nonvolatile product. The solvent selected is one that reacts with the pollutant in an irreversible fashion—for example, the reactions of ammonia and carbon dioxide gases with acidic and basic solvents, respectively. These reactions produce carbonic acid (H_2CO_3) and ammonium hydroxide (NH_4OH). The solubilities of these acids and bases are much greater than gaseous CO_2 or NH_3. Primary factors affecting the choice of an absorbent in chemical absorption are the solubility of the pollutant, reactive properties of pollutant and absorbent and the subsequent analytical method to be used. Care should be taken to avoid an absorbent that will interfere with subsequent chemical analysis.

A typical process involving chemical absorption is the reaction of SO_2 and aqueous H_2O_2 to produce sulfuric acid. The concentration of SO_2 is determined by titrating the H_2SO_4 formed with $Ba(ClO_4)_2$. This procedure is currently the reference method for determining SO_2 emissions from stationary sources.

TABLE 1.1. Solubility of Selected Gases in Distilled Water at 20°C.

Gas	Volume Absorbed per Volume of Water*
Nitrogen	0.015
Oxygen	0.031
Nitric oxide	0.047
Carbon dioxide	0.878
Hydrogen sulfide	2.582
Sulfur dioxide	39.374

*Gas volumes reduced to 0°C and 760 mm Hg.

Collection Efficiency

Each absorption sampling device must be assembled from units found to be most suitable for the specific pollutant involved. It is not necessary to have 100% collection efficiency; however, the efficiency under sampling conditions should be known and reproducible. In some circumstances a sampling system having a relatively low collection efficiency (e.g., 60 to 70%) could be used provided that the desired sensitivity, reproducibility and accuracy are obtainable.

There is much information available in the literature concerning optimum flow rates for specific pollutants and collection efficiencies with respect to the pollutant and absorbent for many sampling devices. However, much more information is needed on the variation of collection efficiency with the rate of sampling concentrations of a variety of compounds and the nature of the collecting medium. For available information on gas-liquid absorption theory and the mathematical treatment of the variables affecting collection efficiency, the reader is referred to the literature (Reference section of this chapter). In the present discussion only a qualitative description of the factors affecting collection efficiency has been attempted.

Factors Affecting Collection Efficiency

The variables affecting the collection efficiency of methods that use absorbers for the collection of gaseous contaminants may be conveniently considered as those associated with the absorber, such as an acceptable flow rate, bubble size and height of the liquid column; the chemical characteristics of the sampling situation, such as the chemical nature and concentration of the pollutant in the air and the absorbing medium, the chemical nature and concentration of the absorbing solution and the reaction rate; and the physical characteristics of the sampling situation, such as temperature, pressure and pollutant solubility.

ABSORBER CHARACTERISTICS

The gas flow rate through the absorber is one of the major factors determining the collection efficiency of an absorber. Absorption collection efficiency varies inversely with the flow rate. An increase in the flow rate through the solution will decrease the probability of adequate gas-liquid contact. In addition, high flow rate increases the possibility of liquid entrainment in the effluent gas. If varying flow rates are used in sampling, a collection efficiency versus flow rate curve should be determined for each absorber and absorber type. All other variables (e.g., temperature, pollutant and absorbent types, etc.) should be held at the desired values.

The collection efficiency of the absorption process for a gas or vapor by chemical absorption or physical absorption depends on the probability of successful collisions of reagent or solvent molecules with gas molecules. For a given concentration of reagent, this probability of collision will depend on the surface area of the gas bubbles, on the length of the column of liquid through which the bubbles must pass and on the rate at which they rise through the liquid. As the volume of individual bubbles decreases, the surface area presented to the liquid increases. Hence, smaller bubbles have a greater possibility of gas transfer into the absorbent phase. For this reason many absorption devices use fritted discs as opposed to injection-type dispersion tubes to achieve a smaller bubble size. However, due to possible surface reactions that can take place at the frit, fritted bubblers may not be appropriate for certain types of sampling (e.g., for ozone). The length of the column of liquid in the absorber is another prime factor affecting the collection efficiency. The longer the gas bubble is in contact with the liquid, the more pollutant is transferred. However, in many cases this variable cannot be used to its maximum advantage; for example, when the sampled pollutant has a low concentration in the atmosphere, it must be collected in a small absorbent volume so that it is in the sensitivity range of the subsequent analytical method to be used. Bubble rise time is a function of bubble size and absorbent height. A compromise is usually reached by having the smallest feasible bubble size combined with the highest absorbent column possible for the particular analysis (see Table 1.2).

CHEMICAL CHARACTERISTICS

The best situation, with respect to collection efficiency, is to choose an absorbent with a very large capacity for absorbing the pollutant without building up appreciable vapor pressure. This can be accomplished by choosing a chemical reagent that reacts with the pollutant in an irreversible fashion—for example, the irreversible reaction that occurs when carbon dioxide is absorbed in a sodium hydroxide solution to form the carbonate ($CO_3^{=}$) ion.

The concentration of the absorbing medium to be used is a function of the expected concentration of the contaminant encountered and the rate of the particular chemical reaction being used. An excess of the reactant in the absorbing solution is preferable to ensure that all the pollutant is collected and that the reaction rate is at a maximum. Ideally the reaction should be instantaneous, since the period of contact between the pollutant and the absorbent is a short one.

Since the rate of reaction is proportional to concentrations of the reacting substances, other variables being equal, the rate of the process falls off as the reaction proceeds. This phenomenon must be compensated for by

TABLE 1.2. **Absorption Sampling Devices.**

Principle of Operation	Devices	Capacity (ml)	Sampling Rate (l /min)	Efficiency* (%)	Comment
Simple gas-washing bottles. Gas flows from unrestricted opening into solution. Glass conical or cylindrical shape.	Standard	125–500	1–5	90–100	Bubblers are large. Reduction of sampling rate increases efficiency. Several units in series raise efficiency
	Drechsel	125–500	1–5	90–100	Similar to above
	Fleming	100	1–5	90–100	Difficult to clean
Modified gas-washing bottles	Fritted bubbler	100–500	1–5	95–100	Fritted tubes available for simple gas washing, items above. Smaller bubblers provide increased gas-liquid contact
	Glass bead bubbler	100–500	1–5	90–100	Provides for longer gas-liquid contact smaller bubbles
Large bubbler traverses path extended by spiral glass insert	Fisher Milligan bottle	275	1–5	90–100	
	Greiner-Friedrichs	100–200	1–5	90–100	Similar to Fisher Milligan
Impingers—designed principally for collection of aerosols. Used for collection of aerosols.	Greenburg-Smith	500	1–5	90–100	Cylindrical shape
Used for collection of gases. Restricted opening. Fritted tubes available which allow use as bubbler.	Midget	100	0.1–0.5	90–100	
Smog bubbler	Fritted bubbler	10–20	1–4	95–100	

*Under optimum conditions of flow rate, absorbing medium, etc., for a particular pollutant.

increasing the concentration of the absorbing liquid, thereby forcing the reaction to approach completion rapidly.

PHYSICAL CHARACTERISTICS

The primary physical characteristics affecting collection efficiency are pressure, temperature and pollutant solubility in the absorbing medium. In many sampling situations, these variables are fixed by ambient conditions.

The solubility of the pollutant in the absorbing medium is related to its partial pressure (by Henry's Law), and the partial pressure of the pollutant in turn is related to its concentration. The net effect considering ideal gas behavior is that an increase in pollutant concentration in the air will result in an increase in pollutant solubility in the liquid. Increased pollutant solubility, other variables being equal, results in a higher collection efficiency.

An increase in temperature enhances chemical reactions but decreases pollutant solubility in the absorbent. In most cases, the net effect is a decrease in collection efficiency with increasing temperature.

Determination of Collection Efficiency

The method of determining collection efficiency will depend on how the results are to be used. If the most accurate values are needed, the best available method for determining collection efficiency should be used. On the other hand, if only approximate values are needed, a less stringent method for determining collection efficiency may be satisfactory. In all cases, collection efficiency should be defined with respect to the method of determination.

The most accurate method of determining the collection efficiency of a particular absorber is by a trial on a synthetic atmosphere, duplicating in every respect the actual sampling conditions. Calibration techniques consist of both dynamic dilution and static dilution systems. In dynamic dilution, a continuous supply of a known pollutant concentration is available that can be sampled, while the static system consists of a container holding a known volume of pollutant of a known concentration. In both of these calibration procedures, the investigator must be assured that the atmosphere being sampled actually contains the pollutant concentration it is believed to contain.

Another method that may be used for collection efficiency calibration is the comparison of the technique of interest to a previously calibrated method. In this technique, the conditions of the calibrated method are imposed on the method of interest. All variables in both methods should be identical, especially with respect to interferences.

Absorption Devices

Wash Bottle

A variety of devices have been used for sampling pollutants from the atmosphere. One of the simplest and most common devices used is an ordinary gas-washing bottle containing the absorbent plus a gas-dispersion tube for introduction of the pollutant into the solution. A typical device of this type is illustrated in Figure 1.16.

Gas flows from the unrestricted opening into the absorbent solution. A variety of absorbers of this type are available. They are usually glass and may be conical or cylindrical in shape. Typical flow rates through the various devices range from 1 to 5 liters per minute.

The majority of other absorption devices used in atmospheric sampling fall into two categories: fritted-glass absorbers and impingers.

Fritted-Glass Absorbers

A great variety of shapes and sizes of these absorbers is being used. A few are illustrated in Figure 1.17.

These units usually provide the most efficient collection of gaseous pollutants. In addition to the commercially available units, homemade devices may be created using normal gas-dispersion tubes. The fritted part of the dispersion tube is readily available in the form of a disc or cylinder

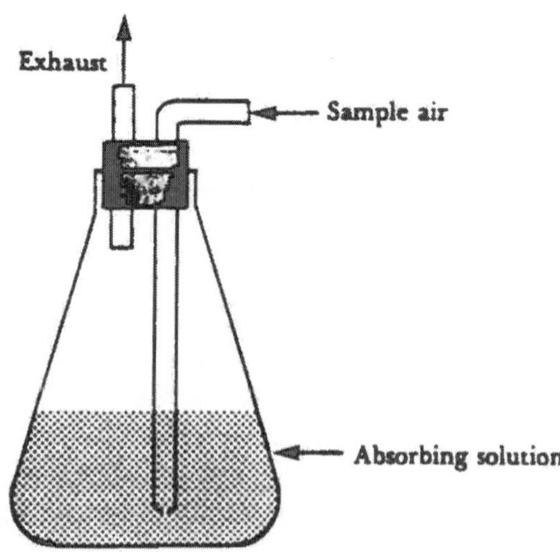

Figure 1.16 Absorption device adapted from an Erlenmeyer flask.

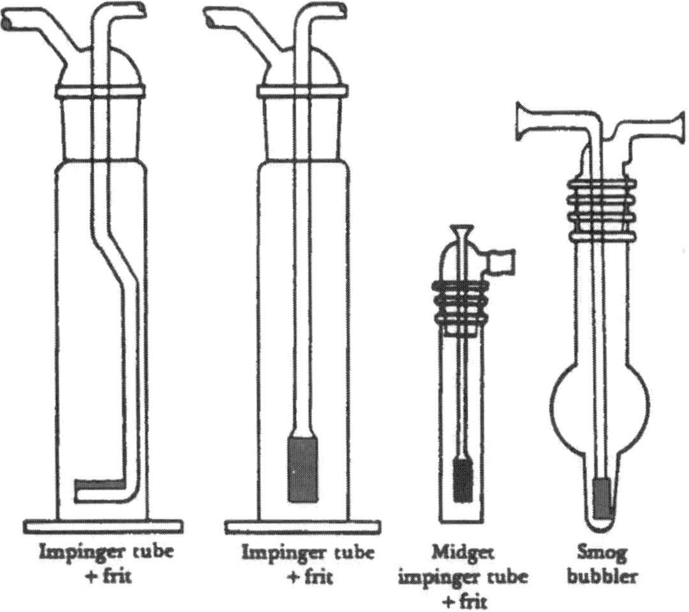

Figure 1.17 Typical fritted-glass absorbers.

of various pore size. The coarse and extra-coarse frits provide good pollutant dispersion with a minimum head loss.

The collection efficiency of any one device will depend on the factors previously mentioned. However, under optimal conditions of flow rate, absorbing medium and pollutant type, many of the fritted-glass absorbers have a collection efficiency in excess of 90%. Several of their more important characteristics are presented in Table 1.2.

Absorbers that use frits with a pore size of approximately 50 micrometers or less gradually become clogged with use. They may be cleaned by surging the appropriate cleaning solution back and forth through the frit and then rinsing with distilled water in the same fashion. Various substances may be removed from the frits by cleaning with the appropriate solvent (e.g., hot hydrochloric acid for dirt, hot concentrated sulfuric acid containing sodium nitrite for organic matter, etc.).

Impingers

Impingers are often used in sampling for gaseous and vaporous pollutants from the atmosphere. Two types of impingers are shown in Figure 1.18.

A limited amount of investigation has indicated that the impinger is

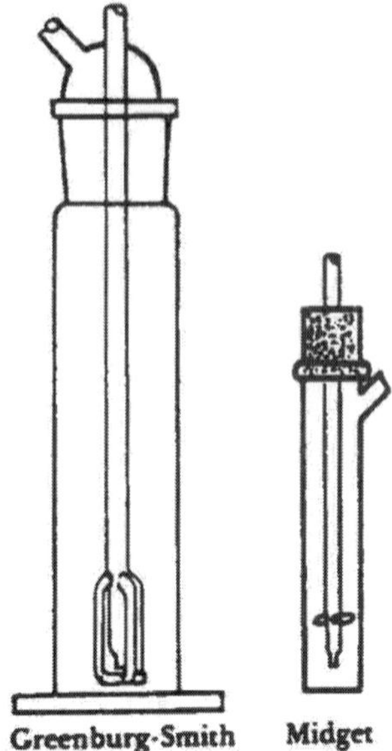

Figure 1.18 Two types of impingers.

somewhat less efficient than the fritted absorber for collecting gaseous pollutants. When several types of absorbers were operated under optimal conditions, the midget impingers were found to be less efficient than the fritted-glass absorber. In addition, the threshold concentration for collection with the midget impinger was found to be somewhat higher than that for several types of fritted-glass absorbers.

STANDARD TEST GAS

INTRODUCTION

Calibration of atmospheric sampling equipment is very important for air monitoring, and it must be done to ensure that the data generated by air monitors represent the actual concentration of pollutants in air. Many factors may affect the calibration of sampling or monitoring devices, preventing them from providing a true measure of the atmospheric contaminant concentrations. The generation of standard test atmospheres is essential to the calibration procedures for continuous air monitoring instrumentation.

There is a need for reliable, accurate methods to generate pollutant gases of known concentration. For example, an air monitor may be designed to operate at a certain efficiency, but due to factors such as reagent deterioration, electrical or electronic component variability, and flow rate changes, data generated by an air monitor may differ from the true concentration of a pollutant in air. To evaluate the performance of sampling equipment, known contaminant concentrations must be introduced; by knowing the input concentration, the output of the monitor can be determined for accuracy and a function generated showing the relationship between the input concentration and output instrument response. The purpose of this section is to discuss the methods of preparing gases of known concentration.

The most important factor in the preparation of these standard atmospheres is devising a method of preparing gases of known concentration. Moreover, in the preparation of the standard atmospheres, devising a method of creating accurate calibration gases at the extremely small concentrations typically found in the atmosphere is often difficult. Many methods are now available for creating gases in high concentrations, but atmospheric testing often requires concentrations in the sub-part-per-million range.

STATIC SYSTEMS

Pressurized Systems

Although pressurized tanks of known contaminant concentrations are usually purchased, they may be prepared by the following procedures. Cylinders containing a pollutant gas are prepared by adding a known volume of pollutant gas and then pressurizing the cylinder with a diluent gas. The gas is then of known concentration and can be used for calibration purposes. The range of concentrations that can be achieved by this method is typically less than 100 ppm to more than 5000 ppm, depending on the stability of the gaseous pollutant. The concentration of the mixture can be calculated as follows:

$$c_{\text{ppm}} = \frac{10^6 \times V_c}{V_d + V_c} = \frac{10^6 p_c}{p_t} \tag{1.10}$$

$$c_{\%} = \frac{10^2 \times V_c}{V_d + V_c} = \frac{10^2 p_c}{p_t} \tag{1.11}$$

where:

c_{ppm} = concentration of gas mixture, ppm by volume
$c_{\%}$ = concentration of gas mixture, percent
V_c = volume of contaminant gas
V_d = volume of diluent gases
p_c = partial pressure of contaminant gas
p_t = total pressure of the gas mixture

Using the rigid container procedure, it is necessary to construct a gas-handling manifold that interconnects the vacuum source, the calibrated volume, the source of contaminant gas and the gas cylinder. Subsequently, the entire system is evacuated; the known calibration volume is flushed and filled at atmospheric pressure with the contaminant gas and isolated, and connecting lines are again evacuated. The contaminant gas is then swept by diluent carrier gas into the cylinder, and the cylinder is pressurized with diluent gas to the desired pressure. Because of compressional heating, the cylinder should be allowed to equilibrate at room temperature before reading the pressure to be used in the concentration calculations. The concentration of the mixture is calculated as follows:

$$c_{\text{ppm}} = \frac{10^6 \times V_c \times p_b}{V_{\text{cyl}} \times p_t} \quad (1.12)$$

where:

c_{ppm} = concentration of mixture, ppm by volume
V_c = volume of pure contaminant gas
V_{cyl} = volume of cylinder
P_b = barometric pressure at time of filling
P_t = final total pressure of cylinder

One factor that must be watched closely when preparing gas mixtures by this method is the thoroughness of the mixing. When introducing the gases into the cylinder one at a time, a layering effect may occur and result in incomplete mixing. This effect can be counteracted by allowing for adequate mixing time before use.

It should be noted that at room temperature and pressure, most gas mixtures conform closely to the ideal gas law. However, at the higher pressures that are present in the cylinders, gaseous mixtures can deviate from this law and create errors of up to 20%. This can be corrected by using a quantity called the compressibility factor (K). The units for the pressure and volume are not important as long as P_c and P_t are in the same units and V_c and V_d are in the same units, since both parts-per-million and percent are unitless quantities.

In commercial practice, compressed gas cylinders of calibration gases are often prepared using high load mass balances; in this procedure, a precise tare weight for an evacuated cylinder is obtained, and the cylinder is weighed again following the addition of the desired trace constituent and after the addition of the diluent gas (prepurified nitrogen in most cases) under pressure. The mass fraction of the contaminant gas is converted to a volumetric concentration by application of the usual formula involving molecular weight and molar volume.

When prepared by the user, a pressure dilution technique of some sort is generally necessary. Here the volume of the contaminant gas introduced, V_c, the volume of the cylinder, $V_{cyl,}$ the evacuation pressure, P_{vac}, and the final total pressure in the cylinder, P_t, must be known.

Using this procedure, the cylinder is evacuated using a vacuum pump capable of producing a very low pressure. Depending on the exact capacity of the pump, the previous contents and size of the cylinder and the final pressure to be used, the evacuation pressure may be ignored in the calculations. For example, if the cylinder was evacuated to a few torrs and the final pressure was 30 atmospheres, the calculation error in ignoring the evacuation pressure (i.e., assuming it is zero) would only be about 1 part in 10,000. Following the evacuation of the cylinder, the contaminant gas is introduced using a syringe technique, as described in the bag procedure, or using a small rigid container of precisely known volume.

As a matter of good practice, cylinders should be continued in the same service and not interchanged; for example, a cylinder formerly used for SO2 span gas should not be converted to NO_2 service. Further, cylinder materials consistent with the gases to be contained therein should be used.

For certain purposes of federally mandated monitoring, the EPA requires that the concentrations of calibration gases contained in compressed gas cylinders be determined by measuring pollutant concentrations generated by the gases using ambient air quality analyzers that have been calibrated using NBS standard reference materials (SRMs) or certified reference materials (CRMs). An EPA-recommended procedure for these determinations, entitled "Traceability Protocol for Establishing True Concentrations of Gases Used for Calibration and Audits of Air Pollution Analyzers (Protocol No. 2)," is contained in EPA 600/4-77-027a *Quality Assurance Handbook for Air Pollution Measurement Systems: Volume II.*

Nonrigid Chambers—Bag Samples

In this procedure, a bag, usually of a flexible, nonreactive plastic material, is filled with a known volume of diluent gas, and a known volume of contaminant gas is added to the system. The diluent gas should be cleaned of all interfering constituents and be nonreactive to the desired contaminant. After allowing for complete mixing in the bag, a sample can be drawn off for calibration purposes. The concentration of contaminant in the bag mixture can be calculated by (if the initial contaminant concentration is 100%):

$$c = \frac{V_c}{V_d + V_c} \tag{1.13}$$

where:

c = concentration of diluted contaminant, ppm by volume
V_c = volume of contaminant gas, μl
V_d = volume of diluent gas, l

One of the first and most important steps in this preparation is the selection of the bag. The bag must be of a flexible material and be chemically inert to the gases it will contain. Chemical inertness is very important; if the contaminant gas reacts with the bag, the amount of contaminant will decrease and the actual concentration of the bag mixture will be unknown. Teflon, Mylar and copolymer Tedlar bags have been widely used because they are inert to most materials. Before any bag is to be used, it should be tested with the contaminant gas to be sure that no reaction will take place between the bag and the gas. After the bag has been selected, it should be checked for leaks, particularly the seams of the bag and the area around the valve, before the sample is introduced. The bag should be flushed and evacuated at least three times to ensure that all unwanted contaminants have been removed. The actual filling of the bag must be done under controlled conditions to guarantee measurement of the volume. Clean air is pumped into the bag through an accurately calibrated flowmeter. The flow rate of the diluent gas must be kept as constant as possible throughout the filling procedure to obtain an accurate measure of volume. After the flow rate is set, a stopwatch is used to get an accurate filling time. The product of the flow rate and the filling time is the total volume of diluent gas added to the system.

The contaminant sample should be introduced into the stream of diluent gas as the bag is filling. This should be done after the bag has filled to about one-quarter of the desired volume. By introducing the sample at this time, mixing can take place as the bag fills to its final volume. The sample is introduced into the diluent gas stream with the use of a syringe. A rubber septum and tee assembly is located in the filling line for insertion of the syringe. Care must be used when handling the syringe to ensure that the desired quantity of contaminant is introduced.

The syringe should have a graduated barrel so that the amount of contaminant entered can be read directly. The plunger of the syringe must be gas-tight so no sample will escape while being injected into the diluent gas stream. A Teflon cap can be fitted over the plunger to eliminate gas escaping during introduction. The sample gas is extracted with the syringe from a source of known concentration. When drawing the sample, the syringe should be filled and evacuated at least six times with the sample gas to eliminate any air that may have been present in the needle. The plunger of the syringe should be drawn well past the desired volume when entering the sample. Just before introducing the sample into the diluent gas stream, the syringe should be adjusted to the desired volume to eliminate any error from air that may have diffused into the needle. Care should be taken in handling the syringe; it should never be held by the barrel of the needle. Heat from the hand of an analyst will cause the gas in the syringe to expand and part of the sample will be lost from the needle. When introducing the sample, the tip of the needle should be in the middle of the diluent gas stream to prevent the sample from being lost on the walls of the filling lines. After depressing the plunger, remove the needle from the gas stream immediately to ensure that none of the sample will diffuse out of the needle and into the gas stream, giving a higher resulting concentration. Mixing time can be decreased by kneading the bag for several minutes.

The concentration of the calibrated gas will change during storage. The decay rate will depend upon the substance being stored, the relative humidity and the bag material. Substances such as nitrogen dioxide and ozone will decay faster than carbon monoxide and hydrocarbons. The decay rate can be decreased if the bag is preconditioned. The preconditioning requires that the bag be flushed several times with the sample gas. The bag should be left overnight at least once with a sample gas in it as a preconditioning step.

It should be noted that the bag-filling method is only an approximate method of preparing a known gas-mixture concentration. A reference method should be used to determine the actual concentration of the bag mixture. Table 1.3 gives the results of carbon monoxide deterioration with time in bags of various materials.

DYNAMIC SYSTEMS

Permeation Systems

The use of permeation techniques for preparation of standard mixtures is very useful for some contaminants. The method is based on the theory that a gas confined above its liquefied form at a constant temperature will permeate through some materials at a constant rate. By putting a liquefied gas into a Teflon tube, for example, permeation of the vapor through the tube will take place because of the concentration gradient that exists between the inner and outer tube walls. By passing different flows of diluent gas over the tube, gases of varying concentration can be generated.

The actual concentration of a sample gas at the EPA's STP (25°C, 760 mm Hg) can be calculated by Equation 1.14.

$$c = \frac{(\mathrm{PR})\left[\frac{24.46}{M}\right]}{Q} \tag{1.14}$$

TABLE 1.3. Carbon Monoxide Sample Deterioration with Time in Bags of Various Materials.

Bag Material:	PVC	Tedlar	Snout*	Aluminized Polyester
No. of bags tested	10	10	5	3
Concn. of calibration gas used to fill bags, ppm	9.0	9.0	8.2	8.2
0 h after filling, av ppm	8.9	8.5	8.2	8.0
Av deviation, ppm	−0.12	−0.5	0	−0.17
Av sqd deviation, ppm	0.056	0.352	0.012	0.030
24 h after filling, av ppm	8.5	7.5	8.0	7.9
Av deviation, ppm	−0.50	−1.5	−0.16	−0.30
Av sqd deviation, ppm	0.306	4.2	0.048	0.097
48 h after filling, av ppm	8.4	6.8	8.3	8.2
Av deviation, ppm	−0.63	−2.2	0.10	−0.03
Av sqd deviation, ppm	0.497	7.7	0.010	0.010
100 h after filling, av ppm	7.9	5.2	8.0	7.7
Av deviation, ppm	−1.2	−3.8	−0.18	−0.50
Av sqd deviation, ppm	1.5	17.8	0.066	0.25

*Consists of layers of polyester, polyvinyl chloride, aluminum, polyamide and polyethylene.

where:

c = concentration, μl/l or ppm by volume
PR = permeation rate, μg/min
Q = total flow rate, corrected to STP, std l/min
M = molecular weight of the permeating gas, μg/μ-mole
24.46 = molar volume of any gas at 25°C and 760 mm Hg, μl/μ-mole

Permeation tubes allow for the generation of gases with concentrations in the sub-part-per-million range.

Permeation tubes are made from a variety of different materials. The material must allow the diffusion of the contaminant gas through the walls and also be inert to the diffusing gas. If some reaction took place between the tube material and the gas, the permeation rate would be affected and might no longer be constant. Teflon, Mylar and Saran Wrap are materials often used because of their chemical inertness and good permeation properties. Before any material is used for the permeation tube, it should be tested to ensure that no changes will occur in the material when it comes in contact with liquefied gas. To test the material, a piece should be placed in some of the liquefied gas it will contain. The material should be removed after a few days and checked to see if any changes in the material have occurred (i.e., brittleness, holes, stickiness, etc.). If there are no apparent changes, the material is probably suitable for use.

The first step in the construction of the tube is to compress the desired sample gas to a liquid state. The liquefied gas is then put into a tube and the tube ends are sealed. One method of sealing the tube ends is to force glass beads or stainless steel balls into the tube ends. To seal properly, these beads should be approximately one and one-half times the inside diameter of the tube. Once the tube has been prepared, it should be stored for at least 36 hours in order to equilibrate. Since the permeation rate is extremely dependent on temperature and relative humidity, the permeation tube should be stored at a constant temperature with dry air or nitrogen passing over it. After equilibration, the permeation rate of the tube can be determined either gravimetrically or by use of a calibrated analyzer.

For the gravimetric determination, the tube should be weighed on an analytical balance and replaced in the storage area. Time required to weigh the tube should be minimized and kept constant to compensate for the effects of moisture absorption. High humidity will cause the permeation tube to absorb moisture, thereby increasing the tube weight. This will yield an erroneously low value for the permeation rate. Absorbed moisture on the tube can form acids that may cause tube blistering, thus changing the permeation rate. The tube should be stored and weighed several times to yield enough data to demonstrate that the permeation rate is constant. The results of these weighings should be plotted on a graph as weight versus time. From the slope of the resulting best "fit" line, the permeation rate can be calculated in micrograms per minute as shown in Figure 1.19.

For certain purposes of federally mandated monitoring, the EPA requires that permeation rate of permeation tubes used in routine monitoring be determined by measuring pollutant concentrations generated by the tubes using ambient air quality analyzers that have been calibrated using NBS standard reference materials (SRMs) or certified reference materials (CRMs). An EPA-recommended procedure for this type of permeation rate determination, entitled "Traceability Protocol for Establishing True Concentrations of Gases Used for Calibration and Audits of Air Pollution Analyzers (Protocol No. 2)," is contained in EPA 600/4-77-027a *Quality Assurance Handbook for Air Pollution Measurement Systems: Volume II.*

Once the tube has been calibrated, it can be used to generate test gases of known concentration. The permeation tube is placed in a stream of diluent gas. The gas passes over the tube and the permeated gas is mixed into the gas stream. The desired concentration can be varied by varying the flow rate of the diluent gas or by varying the permeation tube length. The diluent gas must be kept at a constant temperature during the time the calibration gas is being generated to be sure the permeation rate is constant. To accomplish this, the diluent gas is drawn through a constant temperature chamber before passing over the tube as in Figure 1.20. Permeation tubes are commercially available from many sources offering a variety of precal-

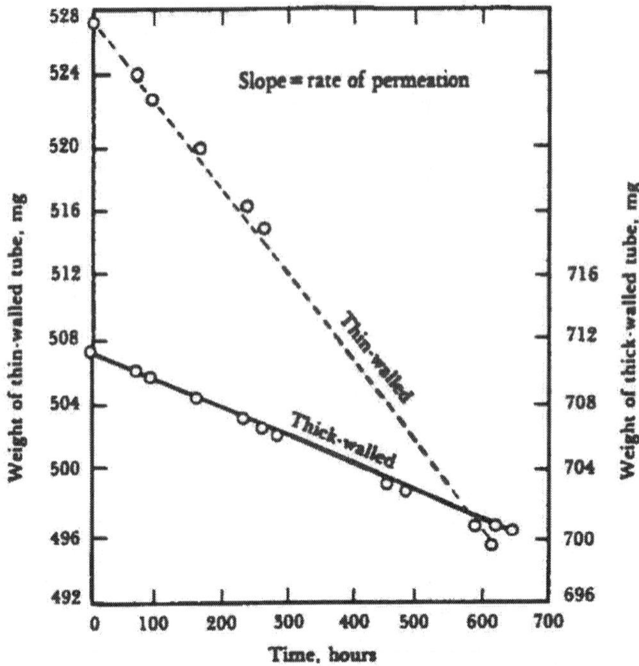

Figure 1.19 Calibration of two permeation tubes.

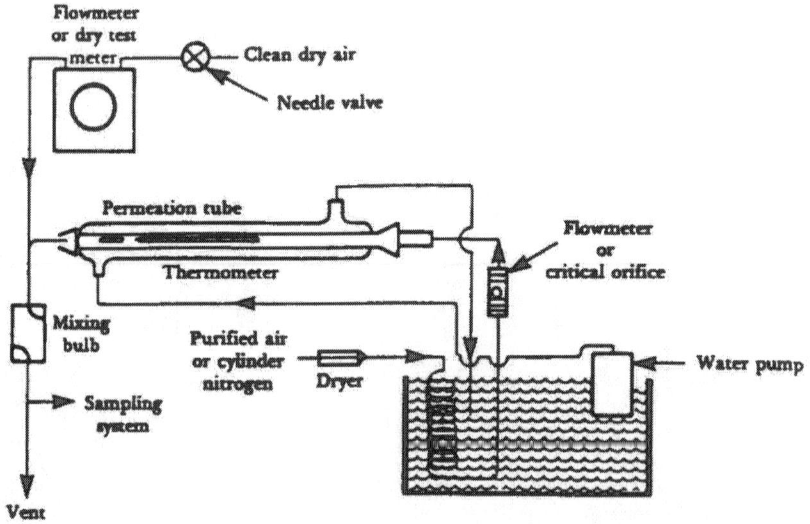

Figure 1.20 Components and flow of a typical permeation system.

ibrated tubes with different permeation rates. The National Bureau of Standards offers some reference sources. Table 1.4 lists permeation rates of a number of compounds through a Teflon film. A number of configurations other than the original tube design are also commercially available. Some of these are designed to provide a longer useful life.

The performance of a permeation device depends on the polymer films used to construct these tubes and the pollutant for which a standard concentration is needed. The important factors to be considered in the use of a permeation device are temperature, humidity, gas stability, equilibration time, etc. These parameters have been studied for nitrogen dioxide, sulfur dioxide, and recently for numerous halogenated hydrocarbons and for permeation tubes constructed with FEP (fluorinated ethylene propylene copolymer) Teflon.

To determine the rate of permeation in this type of device, the tube may simply be removed from the permeation system and weighed to the nearest 0.1 mg on an analytical balance. Generally weighings can be made daily, weekly, or monthly, depending on the gas and type of permeation device. As indicated previously, the permeation rate can be determined by measuring the slope of the least-squares error line used to fit the weight vs. time data. A more rapid calibration, however, can be obtained by using the apparatus shown in Figure 1.21.

TABLE 1.4. Permeation Rates of Some Typical Compounds through FEP Teflon.

Compound	Thickness (in)	Temperature (°C)	Permeation Rate (ng/cm/min)
SO_2	0.012	20 ± 0.5	213
	0.030	20 ± 0.5	138
	0.012	20.1	203
	0.012	29.1	396
NO_2	0.012	13.8	605
	0.012	21.1	1110
	0.012	29.1	2290
Propane	0.012	21.1	53
	0.012	29.1	119
Butane	0.012	15.5	6.4
	0.012	29.1	22.3
CHF_2Cl	0.012	20	2.8
$CF_3CHClBr$	0.016	93	1.3
$CH_2{=}CHCH_3$	0.030	31	0.29
n—C_5H_{12}	0.016	93	0.065
$C_6H_5CH_3$	0.030	20	0.00006

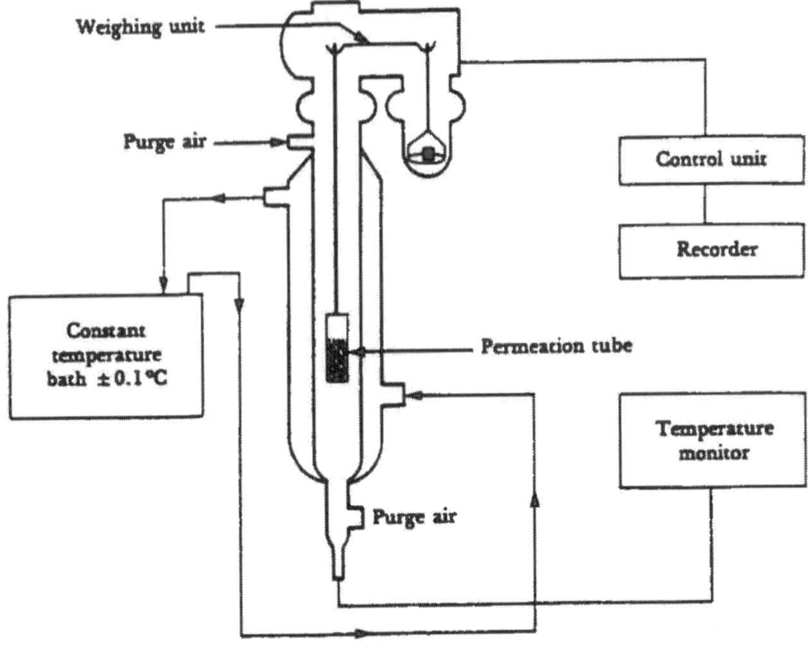

Figure 1.21 Gravimetric calibration apparatus.

PARTICLE SAMPLING

PARTICLE SIZE DISTRIBUTION

In this experiment, bulk density, specific gravity and three measurements of the particle size distribution of a dust sample will be made. The experiments using different types of equipment will be described separately.

Introduction

No discussions of the collection of aerosol particles can be undertaken without an ability to describe the relevant properties of individual particles or the particles constituting the aerosol as a group. The more important properties of an individual particle are its size, shape and density, and for some applications, its dielectric constant. All particles having a specified size, shape and density constitute a "grade" of particles, to be designated as the *i*th grade. The assembly of particles making up an aerosol may be described in terms of its overall concentration and of the distribution of grades by number-, volume-, or mass-fraction within the sample.

Objectives

(1) To become familiar with various techniques used for particle size measurement

(2) To become familiar with the techniques for measuring the bulk density and specific gravity of a particle sample

Particle Size Distribution

An examination of an aerosol sample by almost any of the usual techniques will give size-distribution information in terms of size intervals or ranges. That is, raw data will be generated in the form of the number of particles, or the total mass of particles, falling within a specified range of sizes. Alternatively the total number (or mass) of particles smaller than (or larger than) a specified size may be given. Data of the first kind may be displayed graphically as a histogram in which the intervals of size-range need not all be equal.

The number or weight values displayed by the raw data may be transformed into (1) fractional or frequency values or (2) cumulative values. This will be developed first in terms of number data, and the corresponding calculations for weight basis will be presented later.

Let

n_i = number of particles observed in ith interval
Σn_i = total number of particles of all sizes observed

Then

(1) *Fractional or frequency values*

$$f_i = \frac{n_i}{\Sigma n_i} = \text{fraction of particles in } i\text{th interval, frequency of occurrence of particles in the } i\text{th interval}$$

$$\Sigma f_i = 1$$

(2) *Cumulative values*

$$F_j = \frac{\sum_{i=1}^{j} n_i}{\sum_{i=1}^{n} n_i} = \sum_{i=1}^{j} f_i$$

= fraction of all particles which are smaller than the largest size in the jth interval, called the "cumulative-less-than" fraction

Grade distribution data given in terms of numbers of particles may be converted into terms of weight of particles (or vice versa) by assuming that the weight of a particle is proportional to the cube of its size. Then either all particles in a sample must be taken to have the same density or the actual density of the different kinds of particles present must be determined and used to convert volume to weight. Plots of frequency or cumulative values may then be defined and constructed just as was done for the number data. The definitions are, assuming all particles to have the same density:

$$g_i = \frac{n_i d_{pi}^3}{\sum_{i=1}^{n} n_i d_{pi}^3}$$ = weight fraction of particles in ith grade having diameter, d_{pi}

$$G_j = \sum_{i=1}^{j} g_i$$ = cumulative—less than fraction by weight up to and including jth grade

Mode, Median and Mean Values

It is useful to describe an aerosol by means of a single particle size which represents in some fashion a typical size of the range of particles included in it.

Mode: The most frequently occurring size

Median: The size which divides the sample into two equal portions. Number median diameter (NMD) is that for which $F = 0.50$. Mass median diameter (MMD) is that for which $G = 0.50$.

Mean: There are a number of different definitions of mean size in common use. The appropriate definition to employ in a given instance usually will depend upon the particular aspect of aerosol behavior which is under consideration. One of the most frequently used is the "Sauter mean diameter" which is defined as

$$D_{1,2} = \frac{\Sigma n_i d_{p_i}^3}{\Sigma n_i d_{p_i}^2} = \frac{\Sigma f_i d_{p_i}^3}{\Sigma f_i d_{p_i}^2} = \frac{1}{\Sigma g_i d_{p_i}}$$

The specific surface Ω = (surface of the aerosol)/(volume of the aerosol) may be estimated from $D_{1,2}$, as:

$$\Omega = \frac{6}{D_{1,2}}$$

Log Normal Probability Distribution Function

The log normal distribution is considered to be the most generally applicable function to describe the aerosols encountered in the ambient air, and to describe many process dusts encountered in emission control problems. In a log normal distribution, the logarithm of the particle characteristic (diameter, surface area, volume, etc.) is assumed to be distributed normally about the mean. Special graph paper is available with one axis transformed by the normal probability density and the other axis expressed on a logarithmic scale. If a log normal frequency distribution is plotted on such paper, the points will fall on a straight line.

BULK DENSITY AND SPECIFIC GRAVITY—EXPERIMENT

Theory

There are three densities: true density of the material, apparent density or particle density and bulk density. Each of these densities are different in that they have different values and are used in different aspects of engineering.

True density of a particle is the weight of the material of which the dust is composed per unit volume. In air pollution work, this density is not very important, since all dust particles contain many irregularities such as cracks, dents, pores, etc.

Apparent density is based on the volume of the actual particles, including the small cracks, voids, and other porous spaces within the particles. Therefore, the apparent density will be somewhat less than the true density of the material of which it is composed. Bulk density is defined as the density expressed in kilograms per cubic meter of an uncompressed dust sample. The bulk density will be much less than either the true or apparent densities, since the volume of voids between the particles is also included. Bulk density is useful in the design of hoppers for collection devices.

The specific gravity of a dust is defined as the weight per unit volume of a dust particle in a dust sample, relative to that of pure water at 4°C, expressed as a decimal fraction. Since the density of pure water at 4°C is

1.0 grams/cc, the value which is actually being determined in the specific gravity test is the apparent density of the particulate sample. The method employs the displacement of a liquid by a known weight of the sample. It is desirable to use a liquid with the following characteristics:

(1) A density less than that of the sample so the sample will not float
(2) A low viscosity
(3) A non-wetting tendency of the liquid to the dust

Apparatus and Reagents

Apparatus

- a constant temperature bath
- Leachatelier flask
- long-stemmed funnel
- small piece of tygon tubing
- thermometer
- drying oven (105 – 110°C)
- desiccator
- small spatula
- 50 ml graduated cylinder
- plastic weighing boat
- brush
- analytical balance
- powder funnel

Reagents

- acetone
- kerosene
- silica gel

Procedure

Bulk density

- Dry a dust sample of approximately 50 gm in a dry oven at 105–110°C for a minimum of 2 hours and cool to room temperature in a dessicator.
- Weigh a clean, dry 50 ml graduated cylinder on the analytical balance. Record the weight as W_1. Be careful to touch the cylinder only with gloved hands after you weigh it.
- Introduce the sample into the graduated cylinder. Place the powder funnel on top of the cylinder and gently drop the sample into the

cylinder until the cylinder is filled to the 50 ml level. Tap the sides of the cylinder gently with the spatula until the sample doesn't settle further, but be careful that you do not begin to compress the sample.

- Read the volume of the sample and record the volume as V.
- With the sample still in the cylinder, weigh it and record the weight as W_2.

Specific gravity

- Prepare a constant temperature bath by filling a large container with about 4 inches of water and allow the filled container to sit for about 24 hours or until the water temperature becomes constant. Prepare the Leachatelier flask by making sure that it is clean and completely dry in the neck of the flask. Drying of the neck can be accomplished in most cases by rinsing the neck with acetone.
- Fill the Leachatelier flask with kerosene to between the 0 and 1 ml mark at the bottom of the flask. The long-stemmed funnel should be used. After all of the kerosene has drained from the funnel, remove the funnel from the flask. The small piece of tygon tubing on the stem of the funnel is to aid in keeping the wet end of the funnel from coming in contact with the sides of the flask when the funnel is removed. Of course, the outside of the funnel should be dry before it is inserted into the flask.
- Immerse the flask in the constant temperature water bath maintained at room temperature. Place a thermometer in the liquid in the flask. When the temperature of the liquid is the same as that of the water bath, record this temperature, remove the thermometer and read the volume on the lower neck. Record the volume as V_1. Be extremely careful when removing the thermometer from the flask prior to reading the liquid volume. Also watch that the walls of the k do not become wet during removal of the thermometer.
- Dry approximately 100 gm of the sample in a drying oven at a temperature of 105 – 110°C. Cool it in a desiccator. Weigh out approximately 50 gm of the dried sample to the nearest milligram. Record the weight as W_s.
- Introduce this weighed sample into the flask, a small amount at a time, with a small spatula. The sample should be introduced in small increments by sprinkling the dust off the spatula. If the sample is added too fast, it may bridge over and not be immersed in the liquid. If the dust bridges over and clogs up, try tilting the flask at a slight angle to try to clear the neck. Gently tapping the side of the flask with the spatula will also help. If everything fails, put the spatula through the clogged portion, trying not to get the upper side

of the flask wet. Each sample run will have varying characteristics which will limit the time required for placing the dust in the flask. Do not be alarmed by the amount of dust required to raise the flask level 1 ml. If the proper amount of sample has been added, the level of liquid will rise to between the 18 ml and 24 ml marks. If not, increase or decrease the size of the sample, but be certain to re-weigh the new or remaining sample. Periodically, the mixture should be stoppered and the flask should be rolled in an inclined position to free any entrapped air from the mixture. Failing to do this often enough is a principal cause of error.

- Immerse the flask in the constant temperature bath and insert the thermometer. When the liquid in the flask is at the same temperature as that originally measured, remove the thermometer and determine the volume of the liquid. Record the volume as V_2.

Calculation

- Bulk density $= \dfrac{W_2 - W_1}{V}$ (gm/cc)

  where:

  W_1 = weight of the empty cylinder (gm)
  W_2 = weight of cylinder with sample (gm)
  V = volume of sample (ml)

- Specific gravity $= \dfrac{W_s}{V_2 - V_1} \; \varrho_{H_2O}$

  where:

  W_s = weight of cool and dry sample (gm)
  V_2 = final volume occupied by sample and liquid (ml)
  V_1 = initial volume occupied by liquid (ml)
  ϱ_{H_2O} = density of water (1 gm/cc at 4°C)

OPTICAL MICROSCOPY—EXPERIMENT

Theory

The advantages of particle sizing by optical microscopy are the ability to work with extremely small sample sizes (on the order of several hundreds of particles) and the capability of qualitative identification of the substance, based on the observed shape, color and morphology of the substance.

Two methods are commonly used in sizing particles:

(1) *Ferret's diameter* is measured by measuring the distance across all of the particles in a sample, all in one direction, regardless of whether this is the largest or the smallest distance as shown in Figure 1.22.

(2) *Martin's diameter* is measured by imagining the particle to be divided into two areas by a straight line (in this case, all lines must be in the same direction), such that the area of the particle above and below the line are equal. The length of the line is then the Martin's diameter, as shown in Figure 1.23.

Equipment

- microscope with an ocular micrometer
- slides and cover slips
- glycerol, oil
- stage micrometer
- microspatula

Procedures

(1) Focusing the microscope
   - With the low power objective in place, turn the coarse adjustment to lower the objective until it stops or until it is a short distance above the cover slip. Do this while looking at the slide and objective from the side and not through the eyepiece.

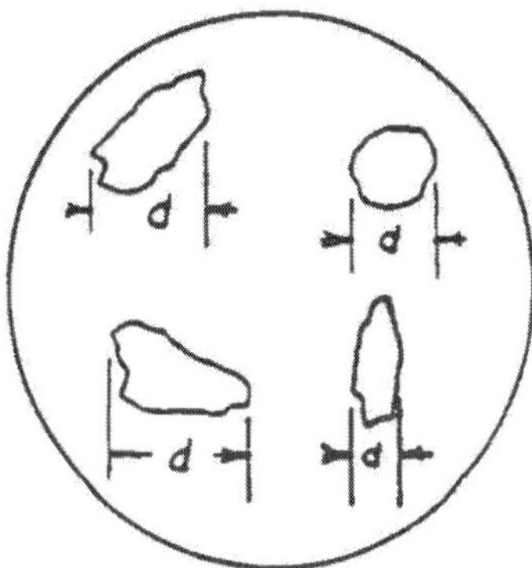

Figure 1.22.

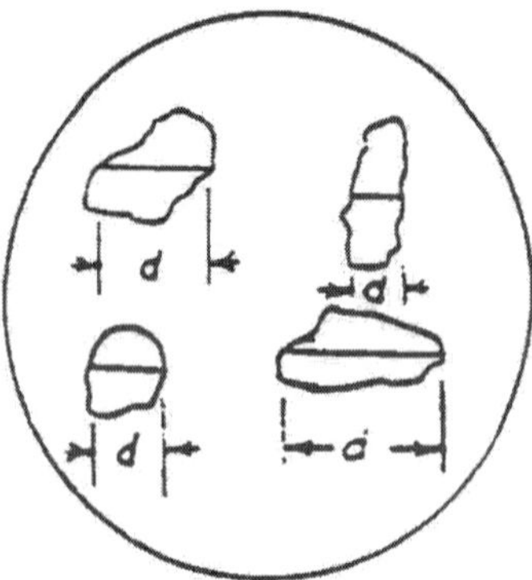

Figure 1.23.

- Look through the eyepiece and slowly focus upward with the coarse adjustment until the specimen comes into view.
- Focus sharply with the fine adjustment.
- Rotate the revolving nosepiece until the high power objective snaps into place. Increase the light intensity to obtain maximum detail in the image.
- Bring the specimen into sharp focus by slight rotation of the fine adjustment knob in one direction or the other, as required.
- If observation with the oil immersion objective is necessary, place a drop of immersion oil on the cover slip. Then revolve the oil immersion (never force the adjustments) into place. The objective should be in the oil, but must not touch the slide.
- Increase the light intensity as required and rotate the fine adjustment knob to obtain a sharp focus of the specimen.
- If this is not successful the first time, repeat the entire procedure.
- Remember: never focus down with your eyes on the eyepiece. Never jam the objective into the slide.

(2) Calibration of the ocular micrometer
- Place a stage micrometer on the object stage of the microscope. The stage micrometer is a microscope slide upon which is mounted a scale 2 mm in length, separated into divisions, the smallest of which is equal to 10 micrometers.
- When a low power objective is used, the length of each ocular division of the scale appears through the eyepiece to be greater than when a high power objective is used.
- To make the calibration, superimpose the ocular scale over the stage scale in such a way as to have at least two lines of both scales coinciding.

- Count the number of ocular divisions that correspond exactly with a given number of stage divisions.
- Determine the number of micrometers for one division of the ocular scale.
- Repeat the same procedure in determining the ocular value when the higher power objectives and the oil immersion objective are used.
- For measuring, the same combination of eyepiece and objective as the one used in calibration must be used.

(3) Preparation of microscope slide

- Place the small amount of dried sample at the center of the microscope slide using the microspatula. (The sample amount may have to be increased or decreased after observing under the microscope.)
- Drop glycerol or other appropriate medium over the particles. (Some samples may be viewed best without glycerol or other dispersing medium.)
- Cover with a cover slip and then slide the cover slip to disperse the particles uniformly. (A rubber policeman or pencil eraser is handy for sliding the cover.)

(4) Particle size measurement

- Place the slide on the microscope stage and focus on it, first using the lower power objective, then rotating to higher magnifications until a good result is obtained.
- Measure the Ferret's diameter of each particle observed beneath the grid one at a time, and record their sizes as an appropriate number of fraction of the eyepiece grid divisions.
- All of the particles in the grid for a selected slide position must be measured and counted before the slide position is changed in order to reduce the statistical bias toward one particular particle size.
- Reposition the slide and measure and count again until about 500 particles in total are counted.

Calculations

(1) Calculate the fractional and cumulative number frequencies of the particles using the definitions given above.

(2) Using the specific gravity measured in Experiment 1, calculate the fractional and cumulative weight frequencies using the definitions given above.

(3) Plot the cumulative number and weight frequencies on log-probability paper. Estimate the median diameters.
(4) Calculate the Sauter mean diameter and the specific surface of the aerosol.

SIZE DISTRIBUTION (BAHCO CLASSIFIER)—EXPERIMENT

Theory

The Bahco Micro Particle Classifier is an elutriator for physically separating dry powders into finely graded fractions. A sample is introduced into a spiral shaped air current flowing towards the center. The spiral current of air has suitable values of tangential and radial velocities, so that a certain part of the sample moves toward the periphery of the whirl and the other part of the sample is carried by the air current towards the center of the whirl. The size, form and weight of the particles determine which direction they will take in the air current. By varying the air flow, it is possible to change the size limit of division; and thus the material can be divided into any number of fractions with limited size ranges.

Equipment

- Bahco Micro Particle Classifier
- brush (1/2–3/4 inches wide, soft bristles)
- drying oven (105–110°C)
- glazed paper or aluminum foil
- desiccator
- spatula
- plastic boats
- balance (0.01 grams –20 grams)
- scale (0.01 grams)
- sieves (60 and 100 mesh)

Procedure

(1) Dry approximately 12 grams of a dust sample in a drying oven at a temperature of 105–110°C to constant weight (about 1 hour). Cool the dust sample to room temperature in a desiccator.
(2) Place the dried dust sample in a tared plastic boat and determine the weight to the nearest 0.01 grams. Record this weight as W_s.
(3) Screen the weighed sample through a 60 and 100 mesh sieve as explained in the later section on "Sieve Analysis."

(4) Make a "blank" run (without sample) on the Bahco to ensure that the unit is free of contaminating sample. This exercise will also familiarize you with the operation of the Bahco prior to running an actual sample. For this purpose, a low number throttle will be used so that a high volume of air will be drawn through the machine (see below).

(5) Using the "blank run" procedure below, run the screened sample from Step 3 above in the Bahco Micro Particle Classifier.

Calculations

Perform the necessary calculations to obtain the cumulative % less than stated diameter. Do this both by mass and by number. Then plot the cumulative number and weight frequency distributions on log-probability paper.

Operation of the No. 6000 Bahco Micro Particle Classifier (Blank Run)

Procedure

(1) Before starting the unit, the operator should disassemble the Bahco into its main constituents. The classifier should be cleaned thoroughly to remove dirt which will contaminate the test samples.

(2) To remove the upper rotor assembly, turn the locking ring counterclockwise to free the lock studs from the ring web. Lift the ring off the studs.

(3) Unscrew the cup 2 or 3 turns for easier grip, and lift the assembly up and out of the lower rotor section.

Caution: When removing and handling the upper rotor assembly, the steel mushroom must not be dented or damaged in any way. This is a carefully machined and balanced surface. If it is bent, it will prevent symmetrical dust movement into the air spiral.

(4) Remove the catch basin by first twisting to disengage it from the restraining bayonet studs.

(5) The parts may be cleaned with carbon tetrachloride to remove any oil or grease. CCl_4 is toxic; use it in the hood.

(6) After cleaning and inspecting, the parts are reassembled by locking each in place.

Caution: The indexing points (x marks) must be matched to preserve rotor balance.

(7) Exceptional care must be taken to assure perfectly clean surfaces before locking the rotor assembly into position. If dirt becomes

embedded in the contact surfaces, it will cause the silting chamber to leak and may throw the upper rotor section out of alignment, producing excessive vibration.

(8) The feeder tip must be screwed firmly into the underside of the feeder hopper.

(9) Close the cover.

(10) Insert the No. 4 horseshoe throttle spacer between the throttle and the stop nut with the curved end of the spacer against the threaded spindle. Push it firmly into position and tighten the throttle down with the fingers. This pressure plus the shape of the spacer will hold it securely in place.

(11) Replace the hand-hole door cover.

(12) Tighten the hopper cover and close the cover.

(13) Start the motor using the switch on the extreme left.

(14) When the motor reaches maximum speed, turn on the feed hopper brush motor and vibrator. Set the vibrator rheostat to maximum intensity.

(15) After a two-minute "blank" run, turn off the feeder switch, turn off the motor switch and gently apply the brake by pushing the knob. There may be some vibrating as the rotor slows down caused by eccentricity from being pushed off center by the brake.

(16) When the rotor stops, raise cover and inspect the rotor, catch basin and fan (Steps 2–4 above). Remove any contaminating particles that may be present and repeat this process until the unit remains clean after the blank run. The sample may now be analyzed as described in the next section.

Operation of the No. 6000 Bahco Micro Particle Classifier (Particle Analysis)

Procedure

Note: Record all data on data sheet, Table 1.5.

(1) Replace the No. 4 throttle spacer with the largest throttle spacer, No. 18, and tighten the throttle down with the fingers.

(2) Replace the hand-hole door cover.

(3) Inspect the rotor assembly to be sure that the catch basin, fan and lock ring are properly positioned with respect to the indexing marks. Tighten the cup in the fan and close the cover.

(4) Open hopper cover. Transfer the dry, weighted test sample carefully into the feed hopper. Close cover.

Caution: Extreme care must be taken to prevent spilling a portion

TABLE 1.5. Particle Size Distribution Bahco Particle Classifier Mod. 6000-209.

Sample Identification______ Date of Analysis______by______ Sample Vessel + Sample (gm) Sample Vessel Tare (gm) Wt. of Entire Sample (gm)				Run No.______ Aliquot for Sizing Run (gm) ______ Screened Retained on 100 Mesh (gm)______ Lost in Screening (gm)______ Wt. of Sample Run in Bahco (gm)______ Tare Wt. of Sample Vessel (gm)______			
	Sample + Tare	Bahco Run Wt. Less than Beginning (gm)	Total Sample Wt. Less than Beginning (gm)	Bahco Run Cum. Wt % Less than	Distribution Cum . Wt % Less than	Equiv. Particle Size at 2.5 Density, μm	Equiv. Particle Size at ____ Density, μm
Beginning		Less Screenings	Incl. Screen & Lost in Screenings	Less Screenings	Incl. Screen & Screenings	Calibration	
Throttle No. 18						1.3	
Throttle No. 17						2.2	
Throttle No. 16						4.2	
Throttle No. 14						7.8	
Throttle No. 12						12.0	
Throttle No. 8						22.5	
Throttle No. 4						29.0	
Throttle No. 0						33.0	
				Total 100 %			
Plus 100 Mesh Screen							
Plus 60 Mesh							

of the sample into the upper rotor assembly. This would contaminate the effluent fraction which is collected in the upper rotor casing.

(5) Start the motor.

(6) When the motor reaches maximum speed, turn on the feed hopper brush motor and vibrator. Set the vibrator rheostat to maximum intensity.

(7) It should take approximately 10 to 15 minutes to feed a 10 gram sample through on the first cut. Feed rates can be regulated by the position of the vibrator rheostat. After 6 minutes, shut off the feed switch and power switch. Check sample in hopper to see if it is about 1/3 to 1/2 gone. If the feed is too fast, cut back on the vibrator rheostat. If too slow, see *Bahco Manual*. Close the lid and continue sampling.

(8) When the sample has passed from the hopper, turn off the feeder switch.

(9) Turn off the motor switch and gently apply the brake by pushing the knob. There will be some vibrating as the rotor slows down caused by eccentricity from being pushed off center by the brake.

(10) When the rotor stops, raise the cover.

(11) Remove lock ring and lift the upper rotor assembly up and out of the classifier using cup. Carry the assembly level to prevent spillage and set it on a piece of glazed paper. The paper should contrast in color with the powder to make complete powder collection easier.

(12) Holding the rotor assembly level, detach the catch basin by rotating counterclockwise and carefully set it aside. Remove the powder adhering to the catch basin lid with a spatula and by brushing it onto the glazed paper with a fine brush. Remove all material from the catch basin, and add it to the material collected from the lid. Any material remaining on the spindle surface should also be collected as residue.

(13) Weigh the residue in a tared plastic boat. Record this weight.

(14) Reassemble the catch basin and the upper rotor assembly and secure the knurled locking ring. Replace the throttle spacer with the next smallest size, No. 16.

(15) Transfer the sample weighed in Step 13 to the feed hopper. Repeat Steps 2 through 13.

(16) Repeat Steps 2 through 15 for each of the remaining throttles, using as feed the residue from the preceding run.

Calculations

(1) Calculate the "Weight Percent Smaller than Indicated Size" for the Bahco run sample only according to the equations:

$$F_{18} = 100(W_p - W_{18})/W_s,$$

$$F_{16} = 100(W_p - W_{16})/W_s,$$

$$F_{12} = 100(W_p - W_{12})/W_s, \text{ etc.}$$

where:

F_{18}, F_{16}, F_{12}, etc. = fraction of sample in weight percent, retained at throttles 18, 16, 12, etc.
W_s = weight of sample
W_p = weight of that portion of the sample passing through the 100 mesh sieve
W_{18}, W_{16}, W_{12}, etc. = weights of sample remaining after fractionation using throttles 18, 16, 12, etc., respectively

(2) Calculate the "Size Distribution of Original Dust–Wt. % Less than Particle Size" by using the equations

$$F_{18} = \frac{100(W_s - W_{18})}{W_s}, \text{ etc.}$$

Note that this distribution includes the screenings and the material remaining on the screen as determined in the blank run procedure.

(3) The columns marked "Equivalent Particle Diameter @ 2.5 Sp. Gr." on the data sheet were determined from a calibration dust.

(4) Use the specific gravity of your dust sample, measured in Experiment 1, to calculate the actual particle diameter, d_{p2} for each throttle setting using the relationship:

$$V_s = \frac{g\rho_p d_p^2}{18n}$$

Therefore:

$$\rho_{p1} d_p^2 l = \rho_{p2} d_{p2}^2$$

(5) Plot the "Actual Particle Diameter" versus "Size Distribution of Original Dust–Wt% Less than Particle Size" and "Actual Particle Diameter" versus "Wt% Less than Indicated Size" from the Bahco run on log-probability paper.

Reference

Instructions for the No. 6000 Bahco Micro Particle Classifier.

SIEVE ANALYSIS—EXPERIMENT

Procedures

(1) Place the weighed sample (W_s) on the 60 mesh sieve, with the 100 mesh sieve underneath, the pan and cover attached.
(2) Place the sieve assembly in the mechanical sieve shaker and tighten into place. Turn on the shaker for approximately 5 minutes.
(3) When the sieving has been completed, remove the cover of the sieve, detach the pan and carefully remove the residue remaining on the 60 mesh sieve to a tared container. Invert the sieve over a piece of glazed paper, and clean the wire cloth by brushing the underside. Add the material thus removed from the wire cloth to the residue removed from the sieve.
(4) Weigh the portion of the material retained on the sieve to the nearest milligram. Record this weight on your data sheet.
(5) Repeat the same procedure as indicated in Steps 3 and 4 for the material remaining on the 100 mesh sieve.
(6) Weigh the material that passed the 100 mesh screen and record this weight.

Calculations

(1) $W_{60} + W_{100} + W_{passed} = W_{sample}$
(2) If the relationship in (1) does not hold, calculate the % dust lost in screening and record on your data sheet.

SAFETY

INTRODUCTION

Researchers, as well as inspection/maintenance personnel, should not be working at a facility unless they have the proper safety equipment, are trained in the use of safety equipment, and are trained to recognize potential safety hazards. *All safety requirements and all agency safety procedures (in case of the regulatory agency inspector) must be satisfied.* This section emphasizes the importance of these safety procedures and provides supplemental information concerning many of the most common procedures.

There are a wide variety of potential hazards which may be encountered. A partial list of these hazards is provided below.

(1) Acute exposure to toxic gases such as SO_2, O_3, NO_2, H_2S and CO due

to entry into confined areas and due to fugitive leaks of flue gas around open measurement ports, corroded door gaskets, and corroded expansion joints and flanges

(2) Exposure to fumes and dust containing asbestos, lead, beryllium and arsenic
(3) Burns due to contact with hot objects such as uninsulated flue gas ducts and portable sampling/measurement equipment
(4) Heat stress due to exposure to hot process equipment, hot flue gas ducts and stacks
(5) Static electrical shock obtained while conducting measurements downstream of electrostatic precipitators or in fiberglass ducts
(6) Falls through roofs or through horizontal surfaces weakened by the accumulation of excess solids or the corrosion of interior supports in air pollution control equipment
(7) Falls on wet walking surfaces in the vicinity of control equipment or on the access ladders to control equipment
(8) Physical injuries due to entrapment in rotating mechanical equipment such as fan sheaves and screw conveyors
(9) Head injuries from falling objects, overhead beams, and protruding equipment
(10) Hearing impairment due to exposure to noise from rappers, compressors, and process equipment
(11) Explosions of fans operating at excessive tip speeds or operating severely out-of-balance
(12) Physical injuries due to the improper opening of access hatches
(13) Asphyxiation due to improper entry into confined areas
(14) Asphyxiation due to free flowing solids discharged from hopper access hatches
(15) Eye injuries due to flying objects or splashing chemicals
(16) Foot injuries due to falling objects
(17) Explosions due to static electricity discharge within ducts being sampled or tested
(18) Exposure to toxic gases, toxic particulate or steam rising from processes below or present in the general vicinity of elevated walkways and platforms
(19) Burns due to contact with hot, free flowing solids
(20) Burns due to contact with high pressure steam leaks

Despite the long list of potential hazards, inspection/maintenance work on air pollution control equipment need not be dangerous. Each individual simply must develop respect for these problems and carefully adhere to the safety procedures adopted by his/her employer.

GENERAL PROCEDURES

Each individual performing work on equipment should adhere to the following two basic rules:

(1) Every situation should be carefully evaluated before work is begun.
(2) Work should be halted if the individual experiences headache, eye or nose irritation, nausea, dizziness, drowsiness, vomiting, loss of coordination, chest pains or shortness of breath. Conditions should be reevaluated before continuing the work.

The first requirement can only be satisfied by conducting the inspection or maintenance work in a methodical manner and at a controlled pace. Undue haste can easily result in careless accidents. It is also important to take a rest break whenever fatigue reaches the point that judgment is affected.

The second rule is necessary, because many of the occupational hazards encountered by inspection/maintenance personnel do not have any easily recognized characteristics such as a distinctive odor. For example, some chemicals such as H_2S, which are easily recognized at low concentrations, cannot be detected at high concentrations; since at high concentrations, they overwhelm and disable the sensory organs. Acute exposure to some materials can occur without any immediate discomfort, then later (within 24 hours) exposure will lead to severe respiratory impairment. For these reasons, it is important that all personnel be familiar with the initial symptoms of acute exposure to possible hazards, so that they are able to leave an area of exposure until environmental conditions can be more fully evaluated.

Each regulatory agency and industrial firm should have certain general safety procedures to minimize the risk involved in equipment maintenance and inspection. These should, at a minimum, consist of the following which are covered in the next sections: Safety Training, a Medical Monitoring Program, and Written Safety Procedures.

SAFETY TRAINING

All personnel should receive regularly scheduled safety instruction in safety procedures, use of personal protective equipment, and recognition of potential physical and inhalation hazards. It is also desirable that each person receive training in first aid and cardio-pulmonary resuscitation. Attendance at the safety training program should be mandatory and should be recorded in each individual's employment file. New employees assigned inspection/maintenance duties should receive this training prior to beginning field work.

If the duties include responding to hazardous chemical spills or fires or the investigation of chemical waste dump sites, considerable additional training may be necessary.

MEDICAL MONITORING PROGRAM

Annual physical examinations should be conducted as a precondition to further work. These examinations should provide screening for evidence of exposure to toxic substances. They need not be a comprehensive health evaluation, since this remains the responsibility of each individual, but should, at a minimum, include the following tests:

(1) Chest X-ray
(2) Electro-cardiogram
(3) Blood tests
(4) Eye examination
(5) Liver and kidney function tests
(6) Pulmonary function tests

WRITTEN SAFETY PROCEDURES

A written procedures manual should be prepared by each regulatory agency and private organization involved in the inspection/maintenance of air pollution control equipment. This should, at a minimum, address use of personal protective equipment, recognition of hazardous conditions, recognition of symptoms of exposure, procedures to be followed in the event of potential hazards, and the reporting of illnesses or accidents.

The manuals should be dated and revised whenever necessary. Each employee should be issued a numbered copy of this manual and should be responsible for adding modifications to the manual.

COMMON HAZARDS

Safe working procedures begin with a full understanding and respect for potential hazards. Those problems most frequently encountered while working on air pollution control equipment are summarized in this section.

Inhalation of Toxic Agents and Asphyxiants

There are a wide variety of toxic agents and asphyxiants which may be accidentally inhaled during the inspection of an industrial facility. *The value of a medical surveillance program and proper safety procedures is demonstrated in the following paragraphs which summarize the potential consequences of acute exposure to some of the most common air contaminants.*

Soluble gases such as sulfur dioxide, ammonia and chlorine have very distinctive odors and can often be detected by taste at very low concentrations. At higher levels, eye, nose and throat irritation occurs. Because of these "warning properties," these types of air contaminants can usually be recognized by inspection/maintenance personnel who can leave the immediate area or put on respirators. If the acute exposure continues, these compounds can cause severe cardio-pulmonary problems including, but not limited to, bronchitis and pneumonia.

Most organic compounds and nitrogen dioxide are not very soluble, and these often penetrate deep into the lower lung. The initial symptoms of exposure are nonspecific and may not be recognized by inspection/maintenance personnel. These symptoms include: dizziness, drowsiness, headache, lightheadedness and nausea. The poor "warning properties" of these types of compounds make them extremely dangerous. Acute exposure may lead to pulmonary edema several hours after the exposure.

The chemical and physical asphyxiants are another group of air contaminants which have very poor "warning properties." Chemical asphyxiants include such gases as carbon monoxide and hydrogen sulfide. Carbon dioxide is the most common physical asphyxiant. Inspection/maintenance personnel may be unable to escape from confined and partially confined areas with high concentrations of these compounds.

The inhalation of particulate matter rarely causes immediate physical discomfort or impairment. For this reason, it is possible for undesirable quantities of toxic materials such as lead, arsenic and asbestos to reach the lung. These can be slowly absorbed into the blood system thereby allowing attack on organs such as the liver and kidneys. In other cases, the materials can lead to severe respiratory problems. Many of the problems resulting from the inhalation of particulate matter develop over an extended period of time and this fact underscores the need for a regular occupational health screening physical.

Table 1.6 is a brief summary of the initial symptoms (if any) of exposure to the common air contaminants. Since inspection/maintenance personnel are usually more at risk from short-term, acute exposures than from chronic exposures, the acute symptoms are emphasized. Table 1.6 does *not* list all materials of significant risk, does *not* list all the possible symptoms, and does

TABLE 1.6. **Symptoms of Exposure to Common Air Contaminants.**

Compound	Route of Entry	Personal Protection Recommended	Symptoms	Comments
Ammonia	Inhalation	Full face mask with ammonia canister, self-contained rebreather or supplied air respirator	Exposure may cause severe eye and throat irritation, nausea, perspiration, vomiting and chest pain.	Bronchitis or pneumonia may result from an intense exposure.
Arsenic	Inhalation and ingestion of dust and fumes	Self-contained rebreather or supplied air respirator	Cough, chest pain, headache, general weakness, followed by gastrointestinal symptoms	Acute arsenical poisoning due to inhalation is rare.
Asbestos	Inhalation of airborne fibers	Dust mask	No symptoms	Prolonged exposure may cause cancer.
Benzene	Inhalation of gas	Self-contained rebreather or supplied air respirator	Eye irritation, headache, dizziness, nausea, drowsiness, convulsions	Acute exposure may lead to unconsciousness and death. Benzene is a confirmed carcinogen.
Beryllium	Inhalation of dust or fumes	Self-contained rebreather or supplied air respirator	Intense, brief exposure may result in nonproductive cough, low grade fever, chest pains, and shortness of breath.	Acute exposure to beryllium can result in chemical pneumonitis with pulmonary edema several hours after exposure.

(continued)

TABLE 1.6. (continued).

Compound	Route of Entry	Personal Protection Recommended	Symptoms	Comments
Cadmium	Inhalation of dust or fumes	Dust mask	Acute exposure may occur without any immediate symptoms. In some cases, there may be dryness in the throat, headache, cough, shortness of breath and vomiting.	Brief exposures to very high concentrations may result in pulmonary edema.
Carbon dioxide	Inhalation of gas	Self-contained rebreather or supplied air respirator	Symptoms are those of asphyxia, including: headache, dizziness, shortness of breath, weakness, drowsiness, and ringing in ears.	Rapid recovery occurs upon return to fresh air.
Carbon monoxide	Inhalation of gas	Self-contained rebreather or supplied air respirator	Initial symptoms include: headache, dizziness, nausea, drowsiness, pale skin color and vomiting.	
Ethylene oxide	Inhalation of gas	Full face respirator and protective clothing	Nausea, vomiting, eye and nose irritation, drowsiness	Acute exposure may lead to unconsciousness and pulmonary edema.
Ethers and epoxy compounds	Inhalation of gas	Self-contained rebreather or supplied air respirator	Eye irritation with tear production, coughing, nausea, drowsiness	

TABLE 1.6. **(continued).**

Compound	Route of Entry	Personal Protection Recommended	Symptoms	Comments
Hydrogen chloride	Inhalation of mist or vapor	Full face mask	Eye and throat irritation	Concentrations of 1000 to 2000 ppm are dangerous even for a brief time.
Hydrogen sulfide	Inhalation of gas	Self-contained rebreather or supplied air respirator	Hydrogen sulfide has a distinctive rotten eggs odor at low concentrations, however, at high levels there may be no odor at all due to olfactory paralysis.	Death may result rapidly from acute exposure. Chemical pneumonia may develop several hours after exposure.
Lead, inorganic	Inhalation of dust, fumes, or vapor; ingestion	Self-contained rebreather or supplied air respirator	No immediate physical symptoms	Work clothes should be washed separately from street clothes. Disposable shoe covers may be advisable. Lead is a cumulative poison.
Mercury, inorganic	Inhalation of dust or vapor	Self-contained rebreather or supplied air respirator	No immediate physical symptoms	Work clothes exposed should not be washed or stored with street clothes. Elemental mercury readily volatilizes at room temperature.

(continued)

TABLE 1.6. (continued).

Compound	Route of Entry	Personal Protection Recommended	Symptoms	Comments
Nickel	Inhalation of dust	Full face respirator	No immediate physical symptoms	Nickel is a suspected carcinogen.
Nickel carbonyl	Inhalation of gas	Full face respirator	Severe exposure can cause headache, dizziness, nausea, vomiting, fever, and difficulty breathing.	Nickel carbonyl is a confirmed carcinogen and may release CO upon decomposing.
Nitrogen oxides (N_2O, NO, NO_2)	Inhalation of gas	Self-contained rebreather or supplied air respirator	Initial symptoms include: cough, chills, fever, headache, nausea, vomiting.	Acute pulmonary edema may follow a 5- to 12-hour period with no apparent symptoms. During exposure, only mild bronchial irritation may be experienced. Concentrations of 100 – 150 ppm are dangerous for periods of 30 to 60 minutes.

TABLE 1.6. (continued).

Compound	Route of Entry	Personal Protection Recommended	Symptoms	Comments
Ozone	Inhalation of gas	Self-contained rebreather or supplied air respirator	Eye irritation, dryness of nose and throat, choking, cough and fatigue	Pulmonary edema may occur several hours after severe exposure.
Sulfur dioxide	Inhalation of gas	Self-contained rebreather or supplied air respirator	Sulfur dioxide has a strong, suffocating odor at levels greater than 0.5 ppm. The taste threshold is 0.3 – 1.0 ppm. Eye irritation begins at approximately 20 ppm.	Survivors of acute exposure may later develop chemical broncho-pneumonia. Concentrations of 400 – 500 ppm are considered immediately dangerous to life.
Sulfuric acid	Inhalation of mist	Full face gas mask	Eye irritation, tickling of the nose and throat, coughing and sneezing. Concentrations of 0.1 – 10 ppm may be unpleasant and concentrations of 10 – 20 ppm may be unbearable.	

(continued)

TABLE 1.6. (continued).

Compound	Route of Entry	Personal Protection Recommended	Symptoms	Comments
Vinyl chloride	Inhalation of vapors	Self-contained rebreather or supplied air respirator	Vinyl chloride has a pleasant, ethereal odor; exposure may cause lightheadedness, nausea, and symptoms similar to mild alcohol intoxication.	Vinyl chloride is a confirmed carcinogen.
Chlorine	Inhalation of vapors	Full face mask	Pungent odor at greater than 3.0 ppm. Throat irritation occurs at approximately 15 ppm. Acute exposure may result in cough and chest pain and eye-throat irritation.	Pulmonary edema may occur several hours after severe exposure.

not provide a complete set of recommendations for personal protection equipment. It is intended simply to emphasize the importance of good safety procedures.

The Occupational Safety and Health Standards regulations have very specific and detailed requirements regarding the use of respirators for protection from exposure to air contaminants. Each individual using respirators should ensure that the requirements are satisfied. Basically this means that each individual should know how to select the proper respirator, should be trained in the use of respirators, should be fitted properly, should understand respirator maintenance requirements, and should be physically able to perform the work while wearing a respirator.

Accidental Falls

Extreme caution is warranted whenever it is necessary to walk across a roof or walkway which has a heavy accumulation of solids. Occasionally, these surfaces have not been designed to support this load plus the additional weight of one or two persons. When crossing such areas, it is prudent to remain close to the well trodden path.

A similar problem may be encountered on the roofs of air pollution control equipment. In this case, corrosion of the interior supports can reduce the load bearing capacity of a roof, thus all should be crossed with caution.

Around wet scrubbers and any other equipment using pumps, there is often a layer of water in the adjacent walkways. In some cases, this layer of water may be partially hidden by a cover of dust or fibrous material. If personnel are not careful, the slick area will be discovered in a most painful manner.

An inevitable chore of inspecting air pollution control equipment is climbing up and down ladders. During wet weather, it is probable that the last person to use a ladder deposited a layer of mud on the foot rungs. Climbing the ladder may be dangerous unless the foot rails are used for the hands (rather than the more common practice of grabbing the side rungs).

Noise

Whenever the noise level is high enough such that normal conversation cannot be heard at a distance of 2 feet, hearing protection is necessary. Speech interference studies have indicated that this noise level is approximately equal to 82 to 88 dBA. For comparison purposes, the OSHA limit for continuous exposure over an 8-hour period is 90 dBA.

During work, noise levels greater than 90 dBA may be encountered around electrostatic precipitator rappers, compressors and blow tubes for pulse jet fabric filters, induced draft fans, and process equipment. Even though the inspection/maintenance personnel will usually not be in the high noise location for the full 8-hour period, the use of hearing protection is: (1) consistent with plant safety policies, (2) comforming with OSHA requirements, and (3) improving communication. The latter point may surprise some who instinctively believe that communication is hindered by earmuffs or plugs. Actually, there is some evidence to demonstrate improved hearing with the use of ear protection when noise levels exceed the 90 dBA range. It is believed that the hearing protection prevents distortion of the ear canal at the high noise levels, while it keeps the speech-to-noise levels at a constant level.

Failure to use hearing protection can lead to gradual hearing loss, especially in the frequency range of 1000 to 4000 Hz.

Unfortunately, there are no immediate symptoms of noise-induced hearing loss. The pain threshold for most individuals is 130 dBA, well above the 90 dBA level which can adversely affect hearing.

There are several basic kinds of hearing protection, including: fixed size ear plugs, malleable ear plugs, and earmuffs. When used properly, any of these can reduce levels of noise 30 to 50 dB in the frequency range of 1000 to 4000 Hz. Obtaining a proper fit with a tight acoustical seal is very important.

Head Protection

Personnel should wear hardhats whenever necessary. The hardhats should satisfy the requirements of ANSI Standards for Protective Helmets, Z89.1-1969. These should protect the individual from falling objects (impact force less than 850 lb-ft) and from collisions with low crossbeams or small equipment suspended from the ceiling.

At least once a month, the hardhat should be checked for possible cracks or a defective suspension. The latter is particularly important, since it is the suspension which provides the impact protection. Hardhats must be washed before checking for cracks.

Eye Protection

Personnel routinely pass through process equipment areas where eye protection is mandatory. Glasses with side shields or flexible goggles should be carried and put on when entering these areas. The eye protection must conform to the requirements of ANSI Standards Z87.1-1968. Practice for Occupational and Educational Eye and Face Protection. Contact lenses should *not* be worn.

Burns

Most of the gas streams of concern to personnel are at relatively high temperatures. For example, the gas streams from cupolas and stoker-fired boilers are often 400–600°F. The ducts transporting this effluent gas stream are often uninsulated, and contact with a duct surface can result in a burn.

Handling pitot tubes and other sampling equipment recently withdrawn from a hot duct is another common cause of burns. Insulated gloves are available for this purpose.

Electrical Shock

High static electrical charges can build up in certain ducts and control systems. When personnel attempt to insert a pitot tube, thermocouple, or other probe into the duct, they may receive a considerable shock. In some cases, the static discharge could initiate a serious explosion. Before inserting any portable test equipment or cleanout rod into a duct and gas stream, the test equipment probe and rod should be grounded. This consists of: (1) a proper electrical bond between the probe/rod and the sidewall of the duct (or control device), and (2) a proper ground connected to the sidewall.

CONFINED AREA ENTRY

A confined area may be defined as an enclosure in which dangerous air contamination cannot be prevented or removed by natural ventilation through openings. Access to the enclosure is usually restricted, so that it is difficult for a worker to escape or to be rescued. Examples of such confined areas include fabric filter compartments, electrostatic precipitator penthouses and electrode sections, wet scrubber internals, mechanical collector internals, and hoppers on all types of air pollution control equipment. Maintenance personnel must enter these areas on regular basis; regulatory agency inspectors should *not* enter confined areas. It is important to note:

> Personnel should be trained in proper confined area entry procedures, should be given the necessary respirators and personal protective equipment, and should be required to perform all work in strict adherence to plant safety procedures and OSHA requirements.

Potential dangers of confined area entry can be grouped into three categories: oxygen deficiency, explosion and exposure to toxic air contaminants. Oxygen deficiency is a very common and insidious hazard, since many of the gas streams treated in air pollution control equipment have oxygen concentrations of only 4−12%. Oxygen levels of less than 19.5% produce detectable physiological changes, while levels less than 16.0% result in rapid disability and death. Unfortunately, the initial sensory symptoms of low oxygen levels are simply a slight difficulty in breathing and ringing in the ears. Maintenance personnel may ignore these symptoms and then rapidly become incapacitated. Prior to entering any confined space, the environment inside should first be tested using a portable oxygen analyzer. The probe should be long enough so that the gas actually *in* the confined area is tested, rather than the air leaking in around the partially opened hatch.

Personnel must be cognizant of the explosion potential inside pollution control equipment. Many combustion sources generate moderately high concentrations of carbon monoxide, especially during upset periods. A combustible gas analyzer should be used to detect the presence of CO and organic vapors. The particulate matter collected in many control devices is also potentially explosive.

Some of the most common toxic air contaminants in the confined areas of air pollution control equipment are H_2S, SO_2, NO_2 and particulate. Maintenance personnel should select and use the proper respirator, based on the initial sampling of the interior environment of each confined area.

Electrostatic Precipitators: A precipitator which has been de-energized for a period up to several days may still have a residual electrostatic charge of 10−20 kilovolts on the electrodes. The electrodes could deliver a fatal shock, unless this charge is first dissipated to ground using a "hot stick." A "hot stick" normally consists of a long wooden stick with a metal hook on one end and a flexible metal conductor connected to a ground. This should be used each time entry is made into a penthouse and electrode zone. Regular maintenance (primarily involving cleaning and testing) is necessary for the hot sticks. If the sound of "buzzing" characteristic of a hive of bees is ever heard upon opening of a hatch, the unit is still energized. Work should be immediately halted!

While working inside an electrostatic precipitator, a securely connected ground should be kept in plain view at all times. If the power supply is accidentally energized this ground could provide some degree of protection.

Fabric Filters: Large fabric filter systems are often built with a number of compartments, one or more of which may be isolated for maintenance. Entry to the off-line compartments for maintenance should be done with the same care used for entry into any other confined area. The dampers which isolate one compartment from another often leak, which can result in high concentrations of toxic gases such as H_2S and CO in the off-line compartment.

While performing work inside a fabric filter, care should be taken that the self-contained rebreather or supplied air respirator does not damage the bags adjacent to the walkways; there is often very little clearance inside fabric filters. Before entering fabric filters, the size of the hatch should be considered with respect to the difficulty in removing a worker in an emergency situation. Personnel inside fabric filters (also electrostatic precipitators and mechanical collectors) are particularly vulnerable to heat stress due to the additional work inherent in the use of respirators and the moderately high temperatures usually found inside this equipment.

The purpose of the preceding discussions is to emphasize the importance of proper entry procedures for air pollution control equipment. Some basic considerations have been introduced. Individuals involved in this work should carefully read OSHA requirements and other references on the subject.

CHAPTER 2

Calibration Procedures and Experiments

CALIBRATION OF FLOW MEASURING DEVICES

FLOW measurement is basic to any sampling scheme. Flow measuring instruments are classified as primary, intermediate or secondary standards, depending on their accuracy and physical characteristics. In these experiments, you will use instruments in each of the three categories of flow measuring techniques, e.g., moving bubble, wet test meter and orifice meter. From this experience, you will learn to use these flow measuring instruments and be better prepared to select the proper flow measuring technique when required in a sampling program for atmospheric pollutants. Note: Also refer to Chapter 1, "Flow Meters" section, for descriptions of these devices.

The objectives are as follows:

(1) To calibrate a wet test meter using a water displacement technique
(2) To calibrate a mass flow meter
(3) To calibrate a rotameter in a bubbler train using a dry gas meter. The researcher will be able to describe the effects that variation in upstream pressure will produce on flow measurements.
(4) To be able to describe how various factors affect orifice flow. These factors include:
    - vacuum
    - upstream resistance
    - directional placement of a common orifice (hypodermic needle)
(5) To construct a pump performance curve

CALIBRATION OF A WET TEST METER—EXPERIMENT

Introduction

Wet test meters are frequently used as a laboratory standard. The time required for equilibration of water temperature and dissolved gases in the water and the bulk and weight of the meter makes it very difficult to use the wet test meter in the field. Wet test meters are very accurate, yet are classified as an intermediate standard, because they cannot be calibrated by physical dimensional measurement. Therefore, wet test meters must be calibrated against a primary standard. An easy-to-use, inexpensive method to calibrate the wet test meter is the displacement bottle technique. A Class A volumetric flask is used to measure the displaced water and is considered the primary standard.

Procedure (Record Data on Table 2.1)

(1) Assemble the aspirator bottle.
   - Attach one end of a piece of gum rubber tubing to the drain of the aspirator bottle [Figure 2.1(a)] and close off the other end of the tubing with a pinch clamp [Figure 2.1(b)].
   - Insert the three-inch piece of glass or plastic tubing into the rubber stopper [Figure 2.1(c)]. Set aside for use in Step 12.

(2) Fill the aspirator bottle with water, and allow the water to equilibrate to room temperature.

(3) Level the wet test meter. Adjust the meter until the bubble is exactly centered in the level [Figure 2.2(a)].

(4) Fill the wet test meter through its funnel with distilled water until the water just covers the pointer [Figure 2.2(b)].

(5) Attach one end of the two-foot length of vacuum tubing to the wet test meter outlet. Attach the other end of this tubing to a needle valve [Figure 2.2(c)]. Attach tubing of the necessary length to the vacuum source and affix the other end to the needle valve. The tubing should fit tightly over all fittings.

(6) With the needle valve closed, turn on the vacuum source.

(7) Open the needle valve to obtain a flow rate of 5 to 10 lpm of air. Allow the air to flow through the wet test meter for one hour. The purpose of this step is to saturate the water with air to prevent any of the measured air from becoming dissolved and lost during the calibration run.

(8) Cut off the vacuum source.

(9) Disconnect the needle valve and vacuum tubing, so that both the inlet and outlet of the wet test meter are at atmospheric pressure.

(10) Check the wet test meter level and adjust if necessary.

(11) If necessary, draw off water through the small petcock or add water as in Step 4, adjusting the water level so the center of the concave meniscus seen in the sight glass just touches the tip of the pointer.

TABLE 2.1. Wet Test Meter Calibration Data Sheet.

Wet test meter serial no.__________ Name__________

Wet test meter *l*/revolution__________ Date__________

Test flask volume, V_{st}__________ *l*

Test number	Barometric pressure P_b (mm Hg)	Aspirator bottle data: Temperature T_r		Saturator data: Manometer reading, P_s		Wet test meter data: Temperature T_m		Wet test meter data: Manometer reading, P_m		Wet test meter data: Meter reading			Corrected volume V_c (*l*)	Error	Correction Factor C.F.
		(°C)	(K)	(mm H_2O)	(mm Hg)	(°C)	(K)	(mm H_2O)	(mm Hg)	Final, V_f (*l*)	Initial, V_i (*l*)	V_m (*l*)			
Average															

Calculations:

T_r = __________ °C + 273 = __________ K

P_s = __________ mm H_2O x 0.0738 = __________ mm Hg

T_m = __________ °C + 273 = __________ K

P_m = __________ mm H_2O x 0.0738 = __________ mm Hg

$$V_c = \left[\frac{P_b - P_m}{P_b - P_s}\right]\left[\frac{T_m}{T_r}\right] V_{st} = \_\_\_\_\_\_\_\_\_\_ \; l$$

Note: If air is pushed through calibration system, P_s and P_m are to be added to P_b.

$$V_m = V_f - V_i$$

$$V_m = \frac{V_{m1} + V_{m2} + V_{m3}}{3}$$

$$V_c = \frac{V_{c1} + V_{c2} + V_{c3}}{3}$$

$$\text{Error} = \frac{V_m - V_c}{V_c}$$

$$\text{C.F.} = \frac{1}{1 + \text{Error}}$$

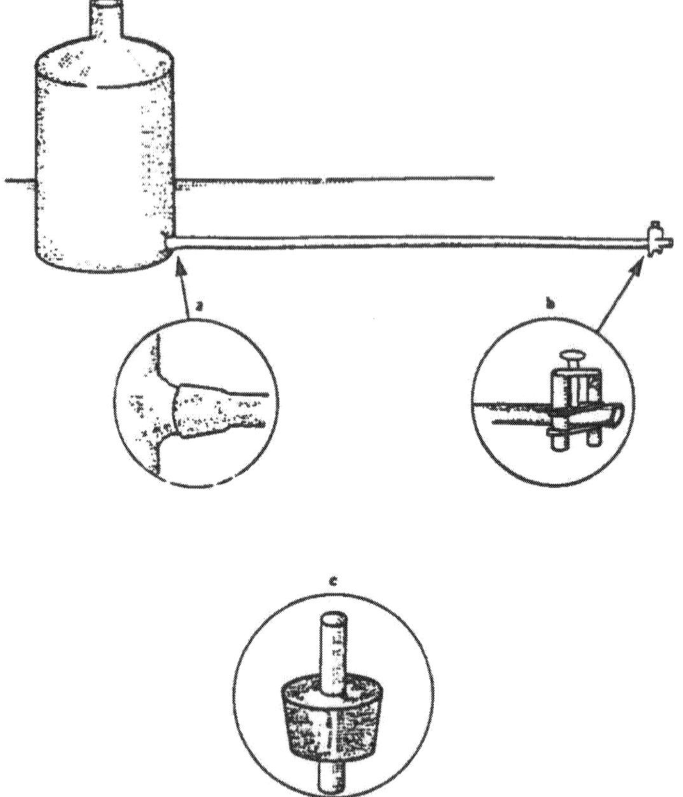

Figure 2.1 Aspirator bottle.

(12) Attach the gum rubber tubing to the glass of plastic piping which was inserted into the rubber stopper in Step 1. Attach the other end to the meter outlet, and assemble the equipment as shown in Figure 2.3. Make certain that the rubber stopper is tight and there are no leaks in the wet test meter by checking the tightness of all fittings.

(13) Open the pinch clamp [Figure 2.3(b)] and allow the aspirator bottle drain hose to fill with water.

(14) Read the initial volume (V_i) from the wet test meter dial and record on the wet test calibration data sheet.

(15) Place a clean, dry 2000 ml volumetric flask (Class A) under the siphon tube, open the pinch clamp and fill the volumetric flask to the 2000 ml mark.

(16) While the water is flowing, record the wet test meter's manometer reading (Δp). Record the following additional data:

- meter water temperature, °C

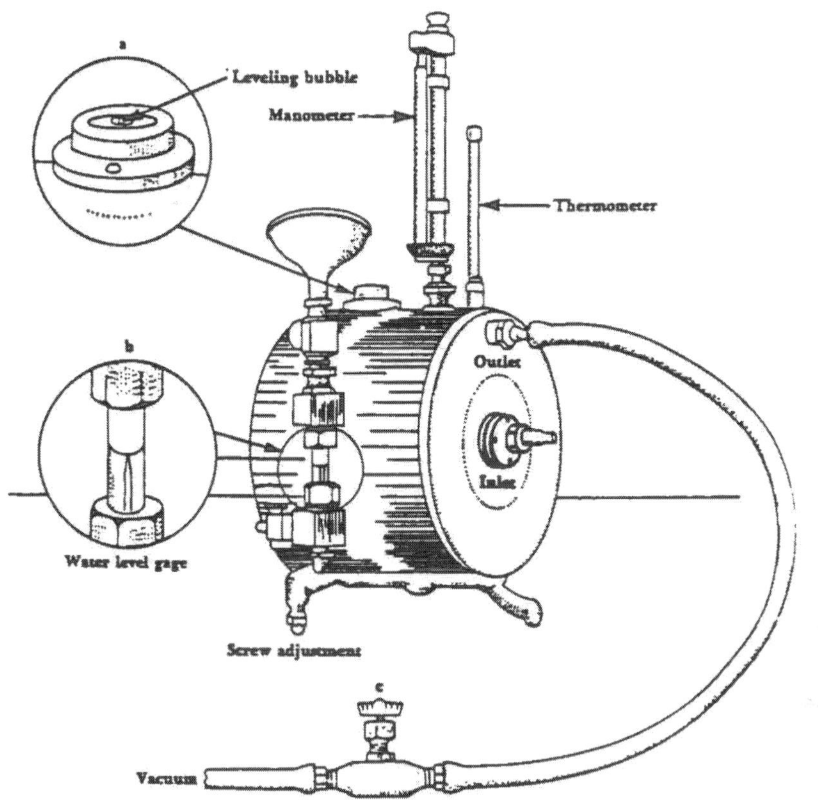

Figure 2.2 Wet test meter.

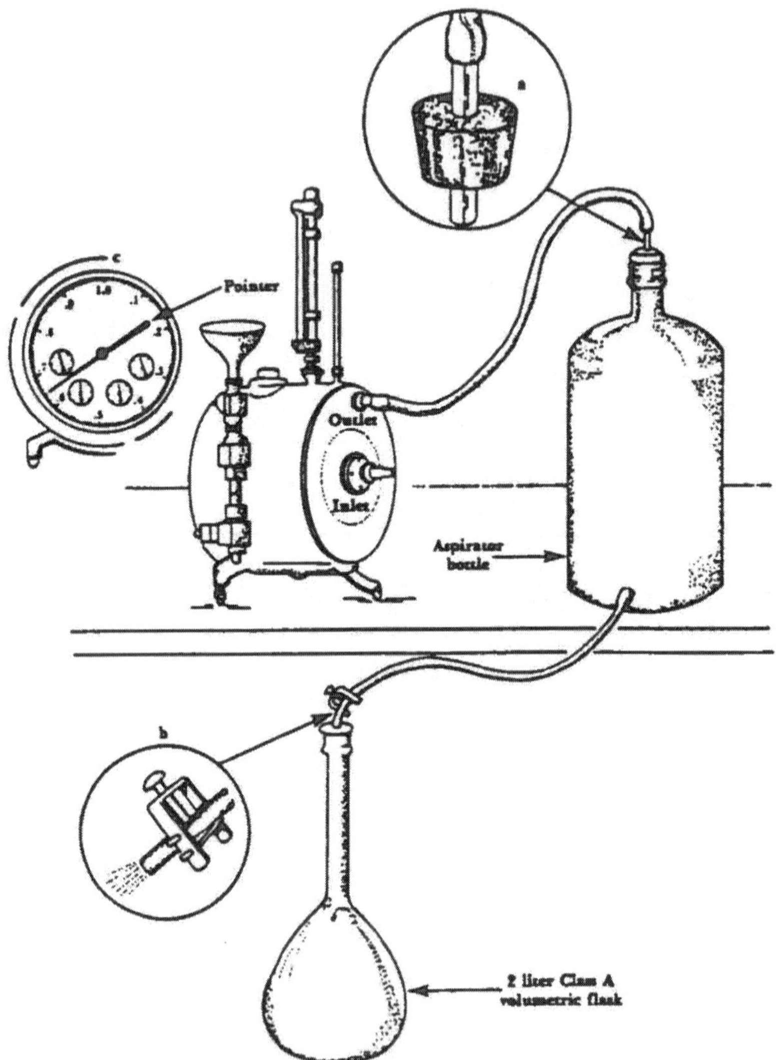

Figure 2.3 Calibration of the wet test meter using aspirator bottle.

- barometric pressure, mm Hg
- volume as indicated on the meter, liters

(17) At the point at which the volumetric flask reaches its calibrated capacity, stop the flow of water by quickly closing the pinch clamp, read the final volume (V_f) from the wet test meter dial, and record this value on the calibration data sheet. Pour the water back into the aspirator bottle.

(18) Repeat Steps 14 through 17 twice.

(19) Calculate the volume at standard conditions, using the following equation for each test:

$$V_s = \frac{P_m}{P_s} \times \frac{T_s}{T_m} \times V_m$$

where:

V_s = volume at standard conditions (25°C, 760 mm Hg)
P_m = barometric pressure corrected for internal meter pressure, mm Hg ($P_b - p$)
T_m = temperature of meter, °K
P_s = barometric pressure at standard conditions, mm Hg
T_s = temperature at standard conditions, °K
V_m = volume as measured by wet test meter, $V_i - V_f$, l

(20) Calculate the average volume at standard conditions by:

$$V_s = \frac{V_{s1} + V_{s2} + V_{s3}}{3}$$

(21) The error can be calculated as follows:

$$\text{Error} = \frac{V_s - V_{st}}{V_{st}}$$

where V_{st} is the volume of the standard flask.

(22) Determine the calibration factor:

$$\text{C.F.} = \frac{1}{1 + \text{Error}}$$

CALIBRATION OF A MASS FLOW METER—EXPERIMENT

Introduction

Mass flow meters operate on the principle that when a gas passes over a heated surface, heat is transferred from the surface to the gas. The amount of electric current required to keep the surface at a constant temperature is a measure of the mass of the gas passing by. Since the amount of heat transferred depends only on the mass and velocity of the gas, these meters measure the true mass flow and have the advantage of measuring flow rates without requiring corrections for changes of temperatures and barometric pressure.

Flow values from these meters are usually given in standard volumetric flow rate, which is the volume occupied by a given mass of the assumed gas being measured at standard temperature and pressure (25°C, 760 mm Hg). Mass flow meters require calibration periodically. It is most convenient to use a primary or intermediate standard flow measuring device such as a wet test meter.

Equipment

- In this experiment, an air velocity meter, which has the same operating principle, is used as a mass flow meter.
- A calibrated wet test meter is used as an intermediate standard flow measuring device.

Procedure (Record Data on Table 2.2)

(1) With the mass flow meter off, adjust the meter to zero with the pointer needle adjustment screw located below the meter face.

TABLE 2.2. Mass Flowmeter Calibration Data Sheet.

Name__________
Date__________
Group No.__________

Wet test meter (WTM) fluid temperature, T_m ______ °C ______ K
Mass flowmeter no. __________
Mass flowmeter range setting __________
Barometric pressure, P_b ______ mm Hg WTM C.F. ______
Transducer no. ______ WTM no. ______

Approximate Mass Flowmeter (% Full Scale)	V_m (l)	ΔP (mm Hg)	Elapsed Time (min)	V_{std} (l)	WTM Q (l/min)	Mass Flow Reading (l/min)
80						
60						
40						
20						

T_{std} = 298 K; P_{std} = 760 mm Hg; mm Hg = 0.0738 × mm H_2O; $P_m = P_b$ mm Hg $-P$ mm Hg; $V_a = V_m \times$ C.F.; $V_{std} = [(P_m)T_{std}/(P_{std})T_m]V_a$; $Q = V_{std}$/time.

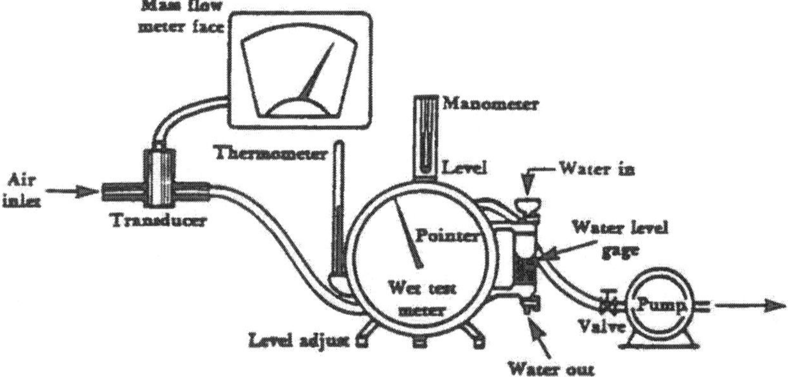

Figure 2.4 Mass flow meter calibration apparatus.

(2) Check the battery level.

(3) Use scale 0–1250 and allow the meter to warm up for one hour.

(4) After the warm-up period, adjust the electronic zero as follows:
- Plug up the inlet and outlet of the transducer head.
- Adjust the meter zero with the pointer needle adjustment screw located below the meter face.
- Unplug the transducer and connect it to the test apparatus as illustrated in Figure 2.4.

(5) Turn the vacuum pump on and adjust the flow rate, with the needle valve, to approximately 80% of full scale of the mass flow meter.

(6) Allow the system to equilibrate for approximately 5 revolutions of the wet test meter's larger pointer.

(7) Record manometer reading (Δp) on wet test in inches of H_2O and convert to mm Hg.

(8) As the wet test meter pointer passes zero, begin timing with a precision stop watch. As the wet test meter pointer passes the three-quarter revolution mark, read and record the mass flow meter reading.

(9) As the wet test meter pointer passes the starting point, stop the stop watch and record elapsed time (Θ) in minutes.

(10) Record the volume of air passing through the wet test meter (V_m) in liters.

(11) Record wet test meter fluid temperature (T_m) in °K, barometric pressure (P_b) in mm Hg, and the vapor pressure of the wet test meter's water (P_v) in mm Hg from Table 2.1.

(12) Calculate actual volume (V_a):

$$V_a = V_m \times \text{C.F.}$$

where C.F. is wet test meter's calibration factor.

(13) Calculate volume at standard condition (V_{as})

$$V_{as} = V_a \times \frac{P_m - P_v}{P_s} \times \frac{T_s}{T_m}$$

where:

$P_s = 760$ mm Hg
$T_s = 298.16°K$
$P_m = P_b - \Delta p$

(14) Calculate flow rate (Q) from V_{as} and Θ

$$Q = \frac{V_{as}}{\Theta}$$

(15) Repeat steps 5 through 14 for mass flow meter settings of approximately 60%, 40% and 80% of full scale.

(16) Plot Q versus mass flow meter readings on linear graph paper.

(17) Construct a best fit curve for the points generated.

CALIBRATION OF A ROTAMETER AT REDUCED PRESSURE—EXPERIMENT

Introduction

Rotameters are widely used as flow measuring devices. Changes in the flow rate causes the float to move up and down the conical tube. Corrections can be made to the rotameter measurements; however, it is simpler and more accurate to calibrate a rotameter at the conditions at which it will be used.

In this experiment, a rotameter will be calibrated with a calibrated dry gas meter at two different upstream pressures.

Procedure (Record Data on Table 2.3)

(1) Record room temperature and barometric pressure (P_b) in mm Hg.

(2) Close Valve B.

(3) Fully open Valve A.

(4) Turn on pump.

(5) Adjust Valve B until the rotameter reads approximately 80% full scale. (Read the middle of the ball.)

TABLE 2.3. Rotameter Calibration Data Sheet.

Room temperature ______________ °C
Mass flowmeter no. ______________
Barometric pressure, P_b ______________
Transducer no. ______________
Rotameter no. ______________

Name ______________
Date ______________
Group no. ______________

Mass Flow Meter Reading	Flow as Indicated by Mass Flow Meter's Calibration Curve (l/min)	Rotameter Reading	Upstream Pressure or Vacuum P_s	Absolute Upstream Pressure P	Pressure Setting No.
					1
//////////	//////////	//////////	//////////	//////////	//////////
					2 reduced pressure

(6) Record the rotameter setting and vacuum pressure reading (p_g).

(7) Record the time (Θ) in minutes, required to pass a known volume of air through the dry gas meter (V_m) in liters by using the dry gas meter's digital volume dial and a stopwatch.

(8) Record dry gas meter air temperature (T_m) in °K.

(9) Calculate actual volume

$$V_a = V_m \times \text{C.F.}$$

where C.F. is the dry gas meter's calibration factor.

(10) Calculate volume at standard condition (V_{as})

$$V_{as} = V_a \times \frac{P_m}{P_s} \times \frac{T_s}{T_m}$$

where:

$P_s = 760$ mm Hg
$T_s = 298.16$ °K
P_m = absolute pressure inside the dry gas meter in mm Hg (usually use the barometric pressure, P_b)

(11) Calculate flow rate (Q) from V_{as} and Θ

$$Q = \frac{V_{as}}{\Theta}$$

(12) Reduce the flow rate to approximately 60% full scale on the rotameter by adjusting Valve B. It is not necessary to set the rotameter at a specific number.

(13) Adjust Valve A so that the pressure reading is the same as Step 7. Repeat Steps 12 and 13 for fine adjustment of the p_g reading (which may require many times). It is important to set the p_g valve at exactly the same reading, but it is not necessary to set the rotameter reading. Just read the rotameter accurately.

(14) Read Steps 7 through 14 for rotameter settings of approximately 40% and 20% of full scale.

(15) Adjust Valves A and B so that a full scale rotameter reading is obtained at a new p_g value which should be 2 to 3 times the former p_g.

(16) Repeat Steps 6 through 15.

(17) Convert gauge pressure readings (p_g) to absolute pressure

$$p_m = p_b + p_g$$

p_g is a vacuum and has a negative sign.

(18) Plot the two resulting curves on the same graph paper. Use the x-axis for flow (l/min) and y-axis for rotameter reading. Note the absolute upstream pressure.

CALIBRATION OF A LIMITING ORIFICE—EXPERIMENT

Introduction

This experiment will also investigate factors affecting flow. Limiting orifices are commonly used to control flow rate. A common device used as a limiting orifice is a hypodermic needle. In this experiment, you will calibrate a hypodermic needle with a soap bubble meter to determine:

- at what vacuum pressure the pressure drop across the hypodermic needle causes the needle to become a limiting orifice
- the effect on flow rate of upstream pressure
- the effect on flow rate of reversing the direction of flow through the needle

Determination of Critical Vacuum (Record Data on Table 2.4)

(1) Make sure that the hypodermic needle is pointed toward vacuum gauge #1a.

(2) Leave needle valve #1 completely open (to open: turn counterclockwise).

(3) Turn on the vacuum pump.

(4) Adjust needle valve #2 until there is about 5 in. of Hg vacuum registered on vacuum gauge #2.

(5) Using the moving bubble meter, determine and record the time required for the specified volume to flow through the bubble meter.

(6) Repeat at other vacuum settings of 10, 13, 15, 17, 19 and 21.

(7) Calculate and correct the flow rate for each vacuum setting, since the use of the soap bubble meter requires correction for temperature, barometric pressure, and the vapor pressure of the soap bubble, which is considered to be that of the vapor pressure of water.

$$Q_{std} = \left\{\frac{\text{Vol.(ml)}}{\Theta\ (\text{min})}\right\} \left\{\frac{P_b - P_v}{760}\right\} \left\{\frac{298.16}{273.16 + T}\right\}$$

where:

Q_{std} = flow rate corrected to standard conditions
P_b = barometric pressure, mm Hg
P_v = vapor pressure of water at room temperature, mm Hg (see Table 2.5)
T = temperature of gas, °C (room temperature)

(8) Plot flow rate on the y-axis and vacuum on the x-axis. At what vacuum pressure does the flow become constant? What is the flow rate?

Effect of Upstream Resistance

(1) Adjust needled valve #2 to maintain 21″ Hg vacuum.

(2) Place resistance in front of the orifice by adjusting needle valve #1, until the vacuum gauge #1 shows 10″ Hg vacuum.

(3) Determine the flow rate under these conditions using the bubble flow meter. Calculate the flow as in the first part.

(4) Calculate the percent change in the flow rate.

(5) The difference between this flow rate and that previously determined

TABLE 2.4. Limiting Orifice Calibration Data Sheet.

P_b ________________ mm Hg
T ________________ °C
Needle gauge ________________

Name ________________
Date ________________
Group no. ________________

Part I Data

Vacuum (in. Hg)	Volume (ml)	Time (min)	Flow Rate, Q_{std} (ml/min)
5	200		
10	200		
13	400		
15	400		
17	400		
19	400		
21	400		
23	400		
25	400		
Maximum vacuum	400		

Part II Data

Condition	Vacuum Gauge #1	Vacuum Gauge #2	Volume (ml)	Time (min)	Flow Rate, Q_{std} (ml/min)
Data from Part I	0	25	400		
Resistance in system	10	25	400		

% Change = ______

Part III Data

Condition	Vacuum (in. Hg)	Volume (ml)	Time (min)	Flow Rate, Q_{std} (ml/min)
Data from Part I	25	400		
Needle reversed	25	400		

% Change = ______

TABLE 2.5. **Saturation Vapor Pressure over Water (°C, mm Hg). Values for Fractional Degree between 50 and 89 Were Obtained by Interpolation.**

Temp. °C	0.0	0.2	0.4	0.6	0.8
-15	1.436	1.414	1.390	1.368	1.345
-14	1.560	1.534	1.511	1.485	1.460
-13	1.691	1.665	1.637	1.611	1.585
-12	1.834	1.804	1.776	1.748	1.720
-11	1.987	1.955	1.924	1.893	1.863
-10	2.149	2.116	2.084	2.050	2.018
-9	2.326	2.289	2.254	2.219	2.184
-8	2.514	2.475	2.437	2.399	2.362
-7	2.715	2.674	2.633	2.593	2.553
-6	2.931	2.887	2.843	2.800	2.757
-5	3.163	3.115	3.069	3.022	2.976
-4	3.410	3.359	3.309	3.259	3.211
-3	3.673	3.620	3.567	3.514	3.461
-2	3.956	3.898	3.841	3.785	3.730
-1	4.258	4.196	4.135	4.075	4.016
-0	4.579	4.513	4.448	4.385	4.320
0	4.579	4.647	4.715	4.785	4.855
1	4.926	4.998	5.070	5.144	5.219
2	5.294	5.370	5.447	5.525	5.605
3	5.685	5.766	5.848	5.931	6.015
4	6.101	6.187	6.274	6.363	6.453
5	6.543	6.635	6.728	6.822	6.917
6	7.013	7.111	7.209	7.309	7.411
7	7.513	7.617	7.722	7.828	7.936
8	8.045	8.155	8.267	8.380	8.494
9	8.609	8.727	8.845	8.965	9.086
10	9.209	9.333	9.458	9.585	9.714
11	9.844	9.976	10.109	10.244	10.380
12	10.518	10.658	10.799	10.941	11.085
13	11.231	11.379	11.528	11.680	11.833
14	11.987	12.144	12.302	12.462	12.624
15	12.788	12.953	13.121	13.290	13.461
16	13.634	13.809	13.987	14.166	14.347
17	14.530	14.715	14.903	15.092	15.284
18	15.477	15.673	15.871	16.071	16.272
19	16.477	16.685	16.894	17.105	17.319
20	17.535	17.753	17.974	18.197	18.422
21	18.650	18.880	19.113	19.349	19.587
22	19.827	20.070	20.316	20.565	20.815
23	21.068	21.324	21.583	21.845	22.110
24	22.377	22.648	22.922	23.198	23.476
25	23.756	24.039	24.326	24.617	24.912
26	25.209	25.509	25.812	26.117	26.426
27	26.739	27.055	27.374	27.696	28.021
28	28.349	28.680	29.015	29.354	29.697
29	30.043	30.392	30.745	31.102	31.461
30	31.824	32.191	32.561	32.934	33.312
31	33.695	34.082	34.471	34.864	35.261
32	35.663	36.068	36.477	36.891	37.308
33	37.729	38.155	38.584	39.018	39.457
34	39.898	40.344	40.796	41.251	41.710
35	42.175	42.644	43.117	43.595	44.078
36	44.563	45.054	45.549	46.050	46.556
37	47.067	47.582	48.102	48.627	49.157
38	49.692	50.231	50.774	51.323	51.879
39	52.442	53.009	53.580	54.156	54.737
40	55.324	55.91	56.51	57.11	57.72
41	58.34	58.96	59.58	60.22	60.86
42	61.50	62.14	62.80	63.46	64.12
43	64.80	65.48	66.16	66.86	67.56
44	68.26	68.97	69.69	70.41	71.14
45	71.88	72.62	73.36	74.12	74.88
46	75.65	76.43	77.21	78.00	78.80
47	79.60	80.41	81.23	82.08	82.87
48	83.71	84.56	85.42	86.28	87.14
49	88.02	88.90	89.79	90.69	91.59
50	92.51	93.5	94.4	95.3	96.3
51	97.20	98.2	99.1	100.1	101.1
52	102.09	103.1	104.1	105.1	106.2
53	107.20	108.2	109.3	110.4	111.4
54	112.51	113.6	114.7	115.8	116.9
55	118.04	119.1	120.3	121.5	122.6
56	123.80	125.0	126.2	127.4	128.6
57	129.82	131.0	132.3	133.5	134.7
58	136.08	137.3	138.5	139.9	141.2

(continued)

TABLE 2.5. **(continued).**

Temp. °C	0.0	0.2	0.4	0.6	0.8
59	124.60	143.9	145.2	146.6	148.0
60	149.38	150.7	152.1	153.5	155.0
61	156.43	157.8	159.3	160.8	162.3
62	163.77	165.2	166.8	168.3	169.8
63	171.38	172.9	174.5	176.1	177.7
64	179.31	180.9	182.5	184.2	185.8
65	187.54	189.2	190.9	192.6	194.3
66	196.09	197.8	199.5	201.3	203.1
67	204.96	206.8	208.6	210.5	212.3
68	214.17	216.0	218.0	219.9	221.8
69	223.73	225.7	227.7	229.7	231.7
70	233.7	235.7	237.7	239.7	214.8
71	243.9	246.0	248.2	250.3	252.4
72	254.6	256.8	259.0	261.2	263.4
73	265.7	268.0	270.2	272.6	274.8
74	277.2	279.4	281.8	284.2	286.6
75	289.1	291.5	294.0	296.4	298.8
76	301.4	303.8	306.4	308.9	311.4
77	314.1	316.6	319.2	322.0	324.6
78	327.3	330.0	332.8	335.6	338.2
79	341.0	343.8	346.6	349.4	352.2
80	355.1	358.0	361.0	363.8	366.8
81	369.7	372.6	375.6	378.8	381.8
82	384.9	388.0	391.2	394.4	397.4
83	400.6	403.8	407.0	410.2	413.6
84	416.8	420.2	423.6	426.8	430.2
85	433.6	437.0	440.4	444.0	447.5
86	450.9	454.4	458.0	461.6	465.2
87	468.7	472.4	476.0	479.8	483.4
88	487.1	491.0	494.7	498.5	502.2
89	506.1	510.0	513.9	517.8	521.8
90	525.76	529.77	533.80	537.86	541.95
91	546.05	550.18	554.35	558.53	562.75
92	566.99	571.26	575.55	579.87	584.22
93	588.60	593.00	597.43	601.89	606.38
94	610.90	615.44	620.01	624.61	629.24
95	633.90	638.59	643.30	648.05	652.82
96	657.62	662.45	667.31	672.20	677.12
97	682.07	687.04	692.05	697.10	702.17
98	707.27	712.40	717.56	722.75	727.98
99	733.24	738.53	743.85	749.20	754.58
100	760.00	765.45	770.93	776.44	782.00
101	787.57	793.18	798.82	804.50	810.21

Source: *Handbook of Air Pollution*, U.S. Dept. H.E.W., P.H.S. Publication No. 999-AP-44.

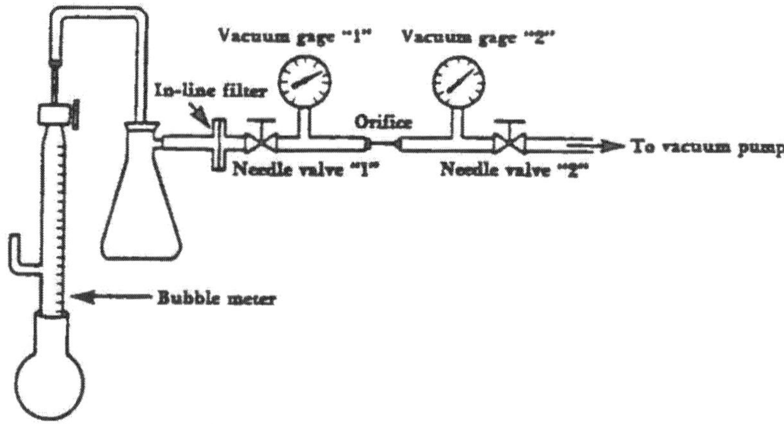

Figure 2.5 Calibration of a limiting orifice using a soap bubble meter.

at the same downstream vacuum (21″ Hg) demonstrates the need to calibrate the hypodermic needle under the same conditions (in the system) that is going to be used in the field.

Effect of Reversing the Needle

(1) Shut off the vacuum pump.
(2) Reverse the needle. The hypodermic needle should now point toward vacuum gauge #2.
(3) Turn on the vacuum pump.
(4) Completely open needle valve #1.
(5) Obtain a vacuum of 21 in. Hg on gauge #2, and determine the flow rate.
(6) Calculate the % change from that determined in the first part at 21 in. Hg vacuum.
(7) This demonstrates the error that would result if you calibrated the needle in one direction and used it in the other direction.

CALIBRATION OF A DRY TEST METER—EXPERIMENT

Introduction

A dry test meter is a positive displacement meter using moving diaphragms or bellows that are alternatively emptied and filled. The movement of the bellows controls the sliding valves and turns the dials to indicate total gas flow. The meter is similar to a consumer dry gas meter for natural gas except that the test meter usually has a dial with smaller graduations for more precise measurements and a digital readout for convenience and minimization errors. They typically can measure a flow rate up to a few tens of liters per minute, but conventional industrial gas meters can be obtained for flows of a few cubic meters per minute or more.

The dry gas test meters can be calibrated to within ± 1% over their range. They can be used as either a laboratory or field instrument for measuring total flows or average flow rates. They are lighter, more rugged, and more versatile than an equivalent wet test meter. Since they are made of metal, the air must be relatively dry and free from corrosive gases to prevent condensation and corrosion that will cause leaks. Because the valving to the bellows can cause slight hesitation in the dial movements, it is preferable to use a minimum of five to ten dial revolutions for any measurements.

Calibration

The dry test meter will be calibrated with a precision wet test meter. To simplify the flow system and calculations, the meters will be set up as shown in Figure 2.6. The saturator humidifies the air to minimize evaporation in the wet test meter. The flow rate is regulated with the valve.

A correction factor for the meter can be defined as

$$C = \frac{\text{actual volume of gas}}{\text{measured volume of gas}} \tag{2.1}$$

where the actual volume is the actual volume of the wet test meter (incorporating its own correction factor) that has been corrected to the pressure and temperature of the dry test meter. The measured volume is the indicated volume of the dry test meter.

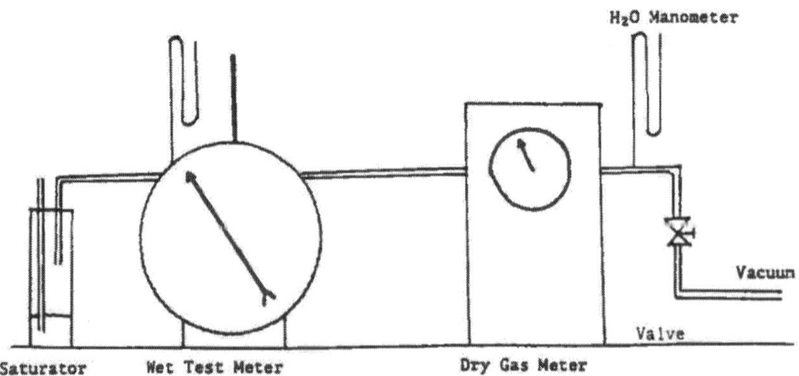

Figure 2.6 Calibration train for the dry gas meter.

Procedure

(1) Read and correct the barometric pressure.
(2) Set up the calibration train including leveling of the wet test meter and adjusting the water level.
(3) Let air flow through the wet test meter for 5 to 10 minutes to equilibrate the water in the meter with the air. Turn off the flow.
(4) Leak check the calibration train.
(5) Begin the flow. Use the valve to set the flow.
(6) At a suitable point, record the readings of the dry gas and wet test meters and simultaneously start a stopwatch.
(7) During the test, record the wet test meter pressure, the dry gas meter pressure and the pressure downstream of the critical orifice. Record the meter temperatures. Note any changes during the test.
(8) After both meters have recorded at least five revolutions, read the meters, stop the stopwatch.
(9) Repeat steps 3–8 to get replicate readings at several flow rates.

Calculations

(1) Correct all pressures from gauge to absolute with the same units (mm Hg or Pa). Correct all temperatures to Kelvin.
(2) For each test, compute the indicated and corrected total flow for the wet test meter.
(3) Correct these wet test meter total flows to the dry gas meter temperature and pressure using the perfect gas laws.
(4) For each test, compute the dry gas meter total flow.
(5) Using the corrected values, compute the correction factor C using Equation 2.1.

CALIBRATION OF A ROTAMETER—EXPERIMENT

Introduction

The rotameter used for this experiment will be a precision rotameter which was calibrated by the manufacturer with nitrogen gas at maximum flow rate of 6 liters/minute. The calibration will be done with a wet test meter using nitrogen to check the manufacturer's scale. The calibration will be repeated using air and helium to determine how well the turbulent rotameter correction formula models the change of volumetric flow rate with calibrating gas. The calibration train is shown in Figure 2.7. The compressed N_2 or He gas flows from a tank through a regulator. The air comes from laboratory supply, through a regulator and a desiccator. Then a needle valve further drops its pressure before passing through the rotameter. The pressure differential across the rotameter and the outlet pressure are measured with manometers. The gas temperature is measured at the rotameter's outlet. The flow then passes through the wet test meter.

Procedures

(1) Read the barometer and correct to determine the absolute pressure.
(2) Set up the calibration train including leveling of the wet test meter and adjusting the water level.
(3) Leak check the system.
(4) Connect the N_2 tank and adjust the regulator to about 20 psi or less.
(5) With the needle valve, adjust the flow to read 6 liters/minute.
(6) With a stopwatch and wet test meter, measure the time for at least 12 liters of gas to pass through the wet test meter.
(7) Repeat step 6 at least three times.
(8) Repeat steps 6 and 7 at a rotameter setting of 3 liters/minute.

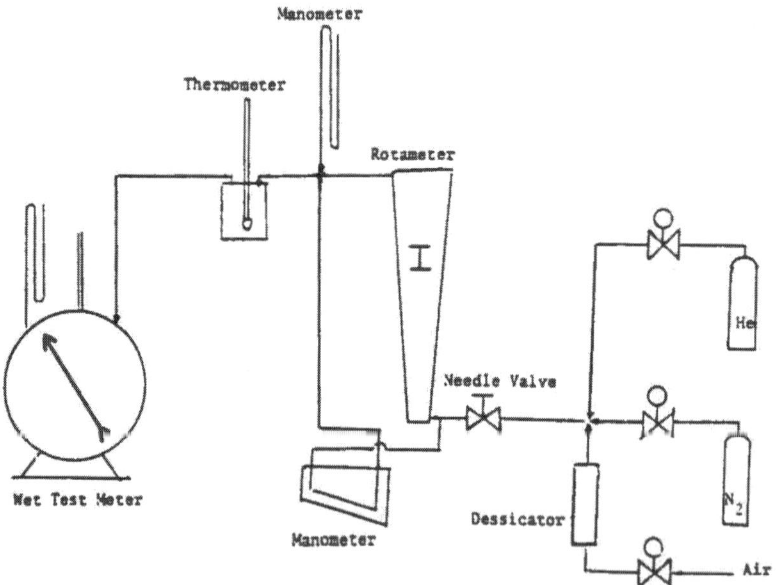

Figure 2.7 Rotameter calibration train.

(9) Replace the N_2 with He tank.
(10) Repeat steps 6–7 (at 6 liters/minute only).
(11) Replace the He tank with the air supply and regulator. Use the desiccator to dry.
(12) Repeat steps 6–7 (at 6 liters/minute).

Calculations

(1) Convert all pressures to absolute values with the same units (mm Hg or Pascals). Convert all temperature to Kelvin.
(2) Convert the timings and wet test meter readings to flow rates.
(3) Since the gases are brought to saturated conditions in the wet test meter, use the sampling correction formula to convert the wet test meter flow rates to the conditions at the outlet of the rotameter.

$$\frac{V_2}{V_1} = \frac{T_2(P_1 - e_1)}{T_1(P_2 - e_2)} \tag{2.2}$$

(4) Use the rotameter correction formula to convert the rotameter reading (liters/minute) to the predicted flow rate at the outlet temperature, outlet pressure and for the gas in question.
(5) For the N_2 data, plot a curve of actual N_2 versus predicted N_2 flow rate at the temperature and outlet pressure.
(6) Use the t-test to estimate if the manufacturer's curve is correct.
(7) For the He and dry air data, make tables of the actual flow rate versus predicted flow rate. If there is a serious discrepancy, please discuss possible reasons.

Calibration of a Small Rotameter with a Bubble Flow Meter

A small flow meter will be calibrated with a bubble flow meter using dry air. The calibration train is shown in Figure 2.8. The lab air supply is passed through a regulator to reduce its pressure and a desiccating tube to remove any moisture. The air passes through a needle valve, the rotameter under calibration and the bubble flow meter.

Procedure

(1) Determine the barometric pressure and correct to determine the absolute pressure.
(2) Set up the calibration train and check for leaks.

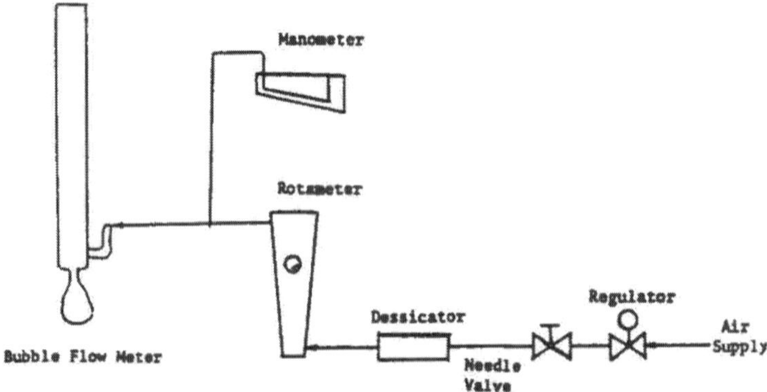

Figure 2.8 Bubble flow meter calibration train.

(3) Set the flow with the needle valve to read the maximum flow with the rotameter. Read downstream pressure.

(4) Squeeze the bulb at the base of the bubble flow meter to produce a bubble that rises up the tube. Continue to send bubbles up the tube until they have wet the tube and will rise to the top.

(5) To measure the flow, time the passage of the bubble up the tube over a known volume. Repeat twice for a total of three measurements.

(6) Reduce the flow to one-half the maximum and repeat step 5.

Calculation

(1) Calculate the actual flow rate through the bubble flow meter.

(2) Plot the actual flow versus indicated flow. Indicate the test conditions.

(3) Since the flow in this rotameter is probably laminar, the rotameter correction formula for turbulent flow should not be used to extrapolate the flow to other temperatures or pressures.

GAS FLOW MODEL STUDIES—EXERCISE (BASED ON IGCI PUBLICATION NO. EP-7)

Introduction

Users of collection control equipment need to be alert to the significant influence of gas distribution on collection efficiency. With time, equipment collection requirements have increased; and suppliers, purchasers and operators are more aware of the influence of gas distribution on collection efficiency. The quality of construction, extent of systems, testing procedures and instrumentation in the art of model studying also have advanced considerably. Therefore, the objectives of this exercise are to establish gas velocity distribution criteria that reflect the best of today's technology and operating experience, to outline techniques, and to recommend procedures for model testing, instrumentation, modeling parameters and flow distribution responsibility.

Decision to Model

During the pre-contract design stage, the layout of the gas inlet and outlet transport flues and the transition connectors to a collection device such as a fabric filter or an E.S.P. should be critically reviewed to ensure that the proposed layout will provide proper gas flow distribution.

Factors influencing a model study are:

(1) *Guaranteed precipitator operating efficiency.* Gas flow uniformity becomes particularly important for operating efficiencies in excess of 99.0% due to:
   - The tendency for the smaller particles to more closely follow the gas flow streamlines.
   - The increased need for almost total suppression of gas bypassing and hopper sweepage. Thus, the higher the guaranteed operating efficiency of a precipitator, the more important the optimization of the gas flow.

(2) *Standardization of Design*–If an installation is a copy of another existing installation, which has satisfactory collection efficiency, the model study may be omitted. If an installation is substantially the same as another, it should be recognized that apparent minor differences in geometry can result in considerable differences in flow distribution.

(3) *Symmetry of Design*–Many large precipitator systems are subdivided into several identical symmetrical parts. In certain arrangements only one of the symmetrical divisions needs to be modeled.

(4) *System Pressure Drop*–If the system pressure loss is guaranteed, the model study will assure minimizing dynamic losses and will locate areas of maximum loss. System losses should be reported as total pressure differences.

Responsibility for Flow Distribution

There should be a definite understanding and agreement between the purchaser and the equipment supplier concerning responsibility for gas distribution as it affects collection efficiency performance guarantees.

A great number of different relationships between the purchaser and the precipitator supplier can and do exist with respect to the design and construction of the flue-diffuser system leading to and away from the precipitator. In general, the precipitator supplier takes responsibility for the gas flow uniformity when the following conditions are met:

(1) The supplier either provides the flue-diffuser design, or has review and veto power over the purchaser's general layout design.

(2) The supplier either provides the final detailed erection drawings, or has review and veto power over the details of the internal flow control devices in the flue- diffuser system. The review may be based on either previous experience, or on the results of a model study.

(3) The supplier either erects the flue-diffuser, or has opportunity to inspect the flues and require any necessary corrections before the installation becomes operational, to ensure that the gas flow control devices have been properly fabricated and installed.

(4) Velocity measurements are made in the completed precipitator to determine whether the gas flow distribution is acceptable.

One important distinction must be recognized. The usual primary guarantee offered by a precipitator supplier is the collection efficiency and outlet emissions. A flow uniformity guarantee, if offered, is usually subordinate to the primary guarantee and comes into play when collection efficiency is deficient by virtue of unsatisfactory flow uniformity.

Gas Flow Uniformity Standards

Using an ESP as the example control device, maximum theoretical precipitator collection efficiency results from perfectly uniform gas velocity. However, due to the inevitable formation of low-speed, viscous, boundary-layer like regions in the collection chamber, and the fact that mechanical and re-entrainment considerations generally require shaped, nonplanar, collecting surfaces, completely uniform flow throughout is neither achievable or desired. Another consideration is that a sheltered low-speed zone is usually deliberately established above the hoppers in order to minimize hopper sweepage.

Nevertheless, gas flow uniformity is desirable to maximize the operating efficiency of the precipitator. The following standards define today's practical limits of gas flow uniformity recommended in precipitator modeling. Depending upon site and design specific circumstances, full-scale flow uniformity can be expected to be somewhat less than that measured in the model.

(1) Within the treatment zone near the inlet and outlet faces of the precipitator collection chamber, the velocity pattern shall have a

minimum of 85% of the velocities not more than 1.15 times the average velocity, and 99% of the velocities not more than 1.40 times the average velocity.

(2) Consideration is often given to having lower than average gas velocity at the upper and lower extremities of the collecting plate to minimize flow over and under the treatment zone. Lower velocity near the bottom of the collecting plate is particularly important to minimize re-entrainment and hopper losses.

(3) For large precipitators subdivided into several chambers, but serving a single source, the uniformity criteria given in 1 above should be considered as a combination of all chambers and evaluated as a single unit.

(4) The individual chamber average velocities should be compared with the overall average velocity to ensure that they do not deviate from it by more than 10%.

(5) Baffles, large structural members and rapping mechanisms can cause dead zones immediately downstream of them. It is not meaningful to include the velocity measurement made in these dead zones with the rest of the velocity data, therefore, these test points may be excluded from the above determinations, provided that all the excluded velocities are less than the average velocity.

Special circumstances such as aerodynamically inferior duct configurations required by limited space availability or the need to prevent dust dropout in horizontal or sloping ductwork when transporting large dust loadings may force the supplier to deviate from the criteria given above. Alternate solutions such as increasing the size of the precipitator are often employed to address these problems.

Velocity Measurement Instrumentation and Procedures

The success of a gas flow modeling effort is dependent on the accuracy with which velocity data is obtained both in the laboratory and in the field.

(1) Laboratory velocity measurement instrumentation should:
- Be reasonably accurate and be repeatable within 2% of the reading or 0.5% of full meter scale. In this service, absolute accuracy is less important than relative accuracy and repeatability. Field instrumentation may sacrifice the criteria above somewhat in exchange for ruggedness.
- Have, for electronic instrumentation, an overall system response time of under one second.
- The system (sensor, signal conditioners, readout/printout conditioner) should be recalibrated frequently as required.

(2) Both laboratory and field velocity measurement procedure should:
- Have a minimum number of test points equal to one-ninth the cross-sectional area of the actual precipitator face (in square feet). To assure proper evaluation of the velocity pattern, a minimum of every third gas passage should be tested. Each passage can then be subdivided into equal points required to meet minimum requirements. However, the vertical test points should not be further apart than 10% of the collecting plate height.
- Preferably have the data taken within three feet downstream of the leading edges of the first bank of collecting plates and within three feet upstream of the trailing edges of the last bank of collecting plates.
- Have either continuous traverses recorded or discrete point measurements taken and recorded.

(3) The two most common types of velocity measurements instruments used are electronic ("hot wire") anemometers and pitot tubes. A hot wire anemometer should have an output signal strength adequate to provide reliable results and must be frequently cleaned of dust contamination in field use. Commonly used electronic anemometers only measure the magnitude of the principal velocity component, and not the direction, or magnitude, of the true velocity vector. The use of streamers, smoke and other qualitative devices are recommended for use in locating flow eddies and recirculating zones.
Draft gauge pitot tube systems may be used instead of electronic anemometers to measure usual duct system velocities, but they are not well suited to measure the low velocities (normally less than 600 FPM) found within precipitator treatment zones.
Two types are commonly used: Prandtl ("L-Head") pitot tubes are capable of higher accuracy ($\pm 1\%$ under ideal conditions), but are susceptible to clogging problems and their use may be precluded in locations with limited access. Stauscheibe ("S-Type") pitot tubes are less accurate than Prandtl ("L-Head") pitot tubes, but are less susceptible to clogging and more versatile in locations with limited access. They also provide slightly higher pressure differences which is helpful when measuring low velocities. It is recommended that the calibration of this type of pitot tube be checked regularly over the full range of velocities.

Modeling Parameters and Techniques

The use of models for velocity distribution studies must be made with a thorough awareness of the laws of fluid mechanics. The dimensionless

Reynolds number defines the ratio of inertial to viscous forces acting on the fluid. Because of the scale factor, the use of full-scale installation velocities results in a Reynolds number of the model being lower than that of the full-size counterpart.

However, as long as the Reynolds number is high enough to ensure that the flow is fully turbulent throughout the model (above approximately 4000), a reliable gas flow model of the precipitator system can be made using 1/16, or larger scale construction. In the flues and diffusers (nozzles), the Reynolds numbers are sufficiently high (typically 1×10^5 to 1×10^6) so that the dynamic losses in the model and full scale are similar. The model flow is unquestionably fully developed turbulent flow.

However, between the precipitator collecting plates, the field Reynolds number can be less than 10,000. Therefore, since it is necessary to maintain turbulent flow (Reynolds number above approximately 4000) in the model, either the distance between parallel collecting plates can be increased, or the velocity through the collection chamber can be increased.

It is important that the collecting plates be represented for each field in the geometric model, since the plates act to preserve existing horizontal distributions. Hence, the absence of one, or several groups of plates could make the modeled outlet velocity distribution different than it is in the actual unit.

In constant cross-sectional area flues, it is not normally critical to model the details of internal structural members, provided that they block less than 5% of the cross section area; and their positioning will not result in their acting as a baffle. However, in the diffuser (nozzle) regions, accurate modeling of structural strutwork is important in obtaining close correspondence between the model and the field.

Modeling Procedure

A geometric gas flow model study has many stages, the omission of any one of which detracts from the overall value of the study:

(1) During the early design stage, the layout of the gas inlet and outlet transport flues and diffusers (nozzles) should be critically reviewed by the precipitator supplier to ensure that the proposed design layout will not lead to unsolvable gas flow uniformity problems. This review procedure should be performed, regardless of whether or not the detailed design and erection of the flues and diffusers will be done by the supplier; since, for the great majority of precipitators, the construction scheduling is such that major changes in the flue layout, at the time of completion of the geometric model study, would cause delays and added expense. The precipitator supplier's previous ex-

perience and gas flow technical expertise will help ensure a good flue and diffuser (nozzle) design.

(2) After duct layout is approved by all parties, model construction can begin. Should the arrangement of equipment and flues require verification or modification, the model construction should be delayed until purchaser' s approval is received.

(3) After checking that the geometric model has been accurately constructed, and is free of leaks, velocity measurements should be made to establish that the flow exiting the inlet flue system and entering the precipitator diffuser (nozzle) is reasonably uniform. If not, alterations should be made in the vaning of the various turns of the transport flue system until satisfactory uniformity is achieved. At this point in the study, the inlet and outlet flue vaning may also be revised in order to minimize the precipitator system's dynamic pressure drop.

(4) Streamers, smoke, dust and other qualitative techniques should be used to locate regions of separated or recirculating flow where particulate dropout could be troublesome.

(5) For multi-chamber precipitators, the velocity data should be examined to see if one, or more, of the chambers is receiving too large a share of the flow. Flow in each chamber should be within ±10% of its theoretical share.

(6) Inlet and outlet velocity traverses in the precipitator collection chamber should be made to determine if the individual velocities are ±20 % of the root mean square average velocity.

Any modifications in the flow distribution devices should be evaluated with another velocity traverse. The sequence of velocity traverse, control device modifications, velocity traverse, etc., should be repeated until an acceptable flow uniformity is achieved.

(7) If field adjustable gas control devices are to be used (i.e., louver-type dampers), then both the maximum and minimum, as well as the optimal settings of the devices, should be established during the model study, in order to establish over what range the device may be fine-tuned in the field without adversely affecting other gas flow factors.

(8) Another qualitative test which should be performed during the geometric model study is the injection of neutral buoyancy smoke into the roof and hopper regions of the collection chamber, particularly near the outlet, to check that significant amounts of gas are not exiting the precipitator without passing through the collecting plates ("sneakage") and that there are not strong flows in the hopper regions (hopper sweepage) which could cause excessive reentrainment.

If the smoke tests indicate that extensive sneakage or hopper sweepage is occurring, then steps should be taken to minimize the effect, since this has a direct harmful impact upon the operating efficiency of the precipitator.

(9) The results of the model study are then incorporated into the final precipitator design drawings.

(10) After field erection of the system is completed, a thorough inspection of flues and diffusers (nozzles) should be made to ensure that the gas flow control and distribution devices have been properly located and installed in the full scale system.

(11) Before the precipitator system is first brought on-line, velocity traverses should be made at the inlet and outlet of the collecting chambers at the same locations that the laboratory tests were made. These measurements serve to establish the degree of agreement between the model and full scale flow pattern.

If field adjustable gas flow devices were included in the precipitator, then velocity measurement should be made before and after each change in their setting.

CHAPTER 3

Ambient Air Sampling

HIGH-VOLUME SAMPLING FOR TOTAL SUSPENDED PARTICULATE MATTER—EXPERIMENT

BACKGROUND

WHEN air pollution control agencies attempt to determine the nature and magnitude of air pollution in their communities and the effectiveness of their control programs, they collect samples of suspended and, sometimes, settleable particulate matter.

Several different sampling techniques and devices – filtration, electrostatic and thermal precipitation, and impaction – may be employed to collect suspended particulate pollutants from ambient air. Of the various techniques, filtration has been found to be the most suitable for routine air sampling. The so-called high-volume (hi-vol) sampler is generally accepted as the instrument of choice for this purpose. Approximately 20,000 hi-vols are operating at federal, state and local air pollution control agencies, industries and research organizations for either routine or intermittent use.

The Environmental Protection Agency has designated the high-volume method as the reference method for total suspended particulate matter (TSP). Certain situations, e.g., sampling for State Implementation Plans (SIPs) and Prevention of Significant Deterioration (PSD) purposes, require the organization responsible for sampling to use the reference high-volume method when determining TSP.

Starting in 1948 as a modified vacuum cleaner, the high-volume sampler has developed as a rugged, reliable sampling system. The sampler and its shelter should be considered as a single, functioning unit (Figure 3.1). The shelter must provide protection for the sampler, and at the same time allow unrestricted access of ambient air from all directions without direct impingement of particles on the filter. A high-volume sampler with a 7- by 9-inch exposed filter area operated in a standard shelter at a sampling flow rate of 39 to 60 cubic feet per minute (1.1 to 1.7 cubic meters per minute) collects particles of up to 25 to 50 micrometers (μm) in aerodynamic

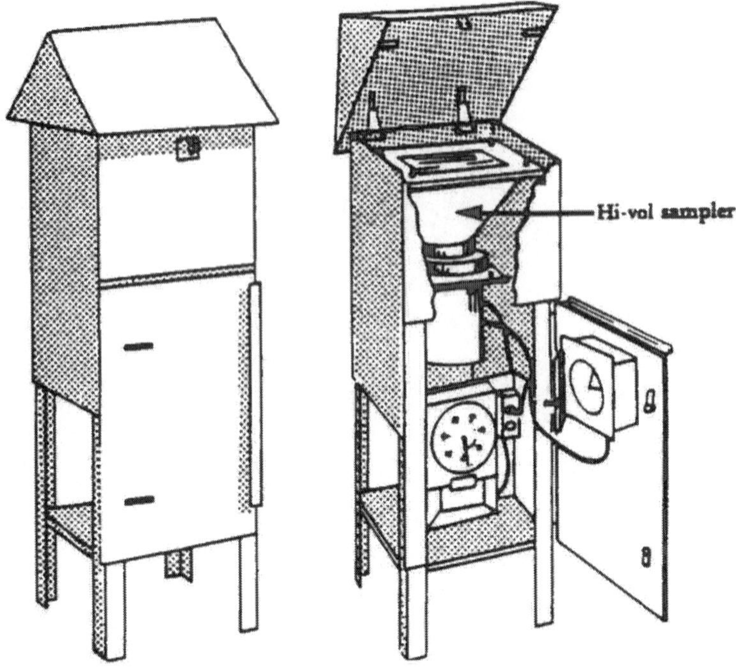

Figure 3.1 Hi-vol sampler in shelter.

diameter, depending on wind speed and direction, and uniformly distributes the sample over the filter surface. The standard peak roof of the shelter, which acts as a plenum above the filter, is placed to provide a total opening area of slightly more than the 63-square-inch exposed filter area, thereby permitting free flow of air into the plenum space (Figure 3.2).

The size of the opening to the filter and the volume of air filtered per unit time will affect the particle size range collected. Distribution of particles on the filter may also be affected. Therefore, any high-volume samplers purchased after February 3, 1983, and used for federally mandated air monitoring, must have uniform sample air inlets that are sized to provide an effective particle capture air velocity of between 20 and 35 cm/s at their recommended sampling flow rates. The particle capture air velocity is determined by dividing the sample air flow rate by the inlet area measured in a horizontal plane at the lower edge of the sampler's roof. Ideally, the inlet area and sampling flow rate of these samplers should be selected to obtain a capture air velocity of 25 $\pm$ 2 cm/s.

FILTER MEDIA

Choice of filter media is influenced by the objectives of the sampling program and the characteristics of the sampler to be employed. Results of a comprehensive study of the characteristics of different types of filter media were published in 1964 by Lockhardt and Patterson. An excellent discussion of filter media and filtration sampling is presented in *Air Sampling Instruments* (American Conference of Governmental Industrial Hygienists 1972).

Glass fiber filters, although not perfect in all respects, have been found to meet most of the requirements for routine particulate matter sampling. Such filters have a collection efficiency of at least 99% for particles having aerodynamic diameters of 0.3 micrometers and larger, low resistance to air flow, and low affinity for moisture, all of which are distinct advantages during sampling. However, in order to eliminate possible weight errors due to small amounts of moisture, both unexposed and exposed filters should be equilibrated between 15° and 30°C with less than ±3°C variation at a relative humidity below 50% with less than ±5% variation for 24 hours before weighing. Figure 3.3 shows the effect of moisture on the weight of glass fiber filters. There would be a similar effect of moisture on the weight of particulate matter.

Samples collected on glass fiber filters are suitable for analysis of a variety of organic pollutants and a large number of inorganic contaminants including trace metals and several nonmetallic substances. Also, glass fiber filters are excellent for monitoring gross radioactivity. However, satisfactory analyses for materials already present in substantial amounts in the filter are

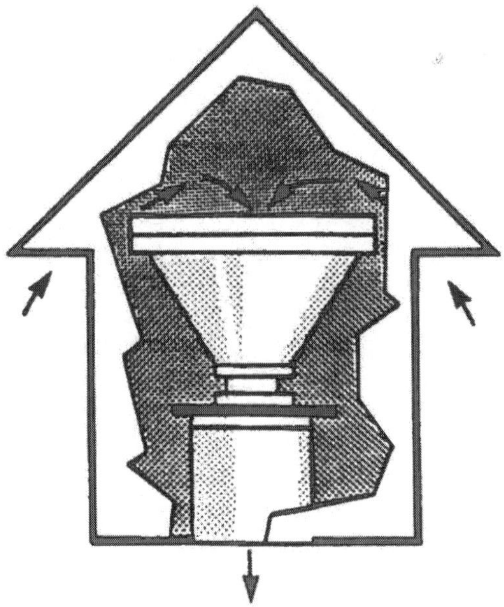

Figure 3.2 Air flow of hi-vol sampler in shelter.

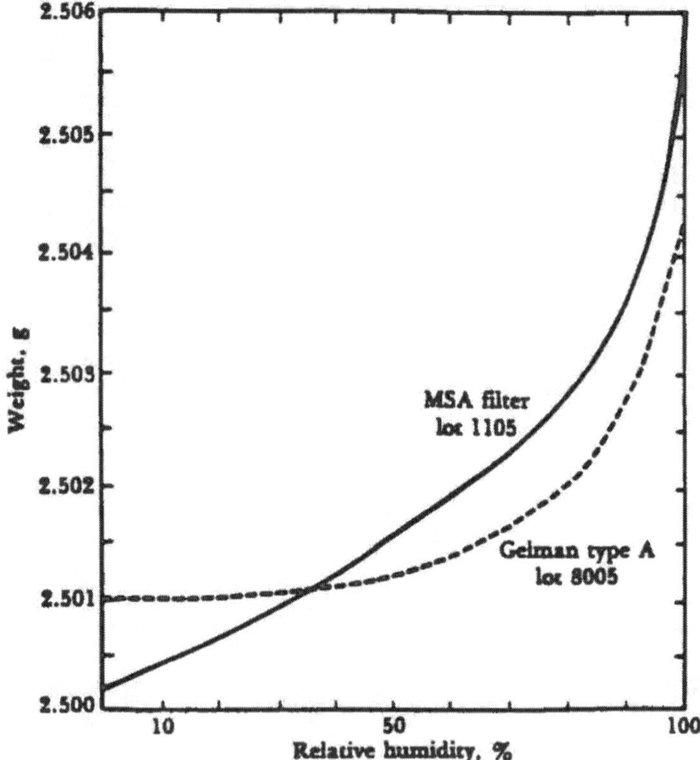

Figure 3.3 The effect of relative humidity on the weight of glass fiber filters at 75°F.

not possible. A random, but statistically significant, sample of new filters should be analyzed to determine whether the filter blank concentration is high enough to interfere with a particular analysis. It is wise to obtain this information before purchasing large numbers of filters to avoid potential problems caused by high blanks.

While glass fiber filter material has been dominant in the measurement of total suspended particulate matter, numerous applications have been found for cellulose filters. Cellulose filters have relatively low metal contents, making them a good choice for metals analysis by neutron activation, atomic absorption, emission spectroscopy, etc. Conventional high-volume samplers usually have to be modified to use cellulose filters because the filters clog rapidly, causing flow to sometimes decrease by as much as a factor of two during a one-day sampling interval. Other disadvantages of cellulose are its irreversible absorption of water and enhanced artifact formation of nitrates and sulfates. These disadvantages can usually be overcome by using a control blank filter. Spectro-quality grade glass fiber filters have sufficiently low background metal contents to make them acceptable for metal analysis, if cellulose cannot be used.

Filters used for federally mandated TSP sampling must meet the specifications listed in Table 3.1. Filters supplied by the USEPA can be assumed to meet these specifications.

USEPA HIGH-VOLUME SAMPLING PROCEDURE

The following procedure is specified in Appendix B of 40 CFR 50 for the sampling of suspended particulate matter.

(1) Number each filter, if not already numbered, near its edge with a unique identification number.
(2) Backlight each filter and inspect for pinholes, particles, and other imperfections; filters with visible imperfections must not be used.
(3) Equilibrate each filter for at least 24 hours.
(4) Following equilibration, weight each filter to the nearest milligram and record this tare weight (W_i) with the filter identification number.
(5) Do not bend or fold the filter before collection of the sample.

TABLE 3.1. USEPA Specifications for TSP Filters.

Parameter	Specification
Size	20.3 ± 0.2 + 25.4 ± 0.2 cm (nominal 8 × 10 in.)
Nominal exposed area	406.5 cm² (63 in.²)
Material	Glass fiber filter or other relatively inert, nonhygroscopic material
Collection efficiency	99% minimum as measured by the DOP test (ASTM-2986) for particles of 0.3 μm aerodynamic diameter
Recommended pressure drop range	42 to 54 mm Hg (5.6 to 7.2 kPa) at a flow rate of 1.5 std m³/min through the nominal exposed area
pH	6 to 10
Integrity	2.4 mg maximum weight loss
Pinholes	None
Tear strength	500 g minimum for 20-mm wide strip cut from filter in weakest dimension (ASTM Test D828-60)
Brittleness	No cracks or material separations after single lengthwide crease

(6) Open the shelter and install a numbered, preweighed filter in the sampler, following the sampler manufacturer's instructions. During inclement weather, precautions must be taken while changing filters to prevent damage to the clean filter and loss of sample from or damage to the exposed filter. Filter cartridges that can be loaded in the laboratory may be used to minimize this problem.

(7) Close the shelter and run the sampler for at least five minutes to establish run-temperature conditions.

(8) Record the flow indicator reading and, if needed, the barometric pressure (P_3) and the ambient temperature (T_3).

(9) Stop the sampler.

(10) Determine the sampler flow rate. If it is outside the acceptable range (1.1 to 1.7 m^3/min [39 to 60 ft^3/min]), use a different filter, or adjust the sampler flow rate.

Warning: Substantial flow adjustments may affect the calibration of the orifice-type flow indicators and may necessitate recalibration.

(11) Record the sampler identification information (filter number, site location or identification number, sample date, and starting time).

(12) Set the timer to start and stop the sampler such that the sampler runs 24 hours, from midnight to midnight (local time).

(13) As soon as practical following the sampling period, run the sampler for at least five minutes to again establish run-temperature conditions.

(14) Record the flow indicator reading and, if needed, the barometric pressure (P_3) and the ambient temperature (T_3).

Note: No on-site pressure or temperature measurements are necessary if the sampler flow indicator does not require pressure or temperature corrections (e.g., a mass flowmeter) or if average barometric pressure and seasonal average temperature for the site are incorporated into the sampler calibration. For individual pressure and temperature corrections, the ambient pressure and temperature can be obtained by on-site measurements or from a nearby weather station. Barometric pressure readings obtained from airports must be station pressure, not corrected to sea level, and may need to be corrected for differences in elevation between the sampler site and the airport. For samplers having flow recorders, but not constant flow controllers, the average temperature and pressure at the site during the sampling period should be estimated from weather bureau or other available data.

(15) Stop the sampler and carefully remove the filter, following the sampler manufacturer's instructions. Touch only the outer edges of the filter.

(16) Fold the filter in half lengthwise so that only surfaces with collected

particulate matter are in contact and place it in the filter holder (glassine envelope or manila folder).

(17) Record the ending time or elapsed time on the filter information record, either from the stop set-point time, from an elapsed time indicator, or from a continuous flow record. The sample period must be 1440 ± 60 min for a valid sample.

(18) Record on the filter information record any other factors, such as meteorological conditions, construction activity, fires or dust storms, etc., that might be pertinent to the measurement. If the sampler is known to be defective, void it at this time.

(19) Equilibrate the exposed filter for at least 24 hours.

(20) Immediately after equilibration, reweigh the filter to the nearest milligram, and record the gross weight with the filter identification number.

(21) Determine the average sampling standard flow rate (Q_{std}) during the sampling period. If the sampler has a continuous flow rate recorder, determine the average indicated flow rate (I) for the sampling period from the recorder trace. Express I with regard to the type of flow rate measuring device used and the method of correcting sample air volumes for ambient temperature and barometric pressure by using the formulas of Table 3.2. Use the resulting value to determine Q_{std} either by referring to a graph of the calibration curve for the sampler's flow rate measuring device or by calculating Q_{std} using the following equation:

$$Q_{std} = \frac{\text{Expressed value of } I - b}{m} \tag{3.1a}$$

where:

Q_{std} = standard volumetric flow rate, std m^3/min
I = indicated flow rate
b = y-intercept of calibration curve for sampler flow rate measurement device
m = slope of calibration curve for sampler flow rate measurement device

If the sampler does not have a continuous flow rate recorder, express the initial and final indicated flow rates (I) with regard to the type of flow rate measuring device used and the method for correcting sample air volumes for ambient temperature and barometric pressure using the formulas of Table 3.2. Use the resulting values to determine the initial and final sampling standard flow rates (Q_{std}) either by referring

TABLE 3.2. **Formulas for Expressing Indicated Sampling Flow Rates.**

Type of Sampler Flow Rate Measuring Device	Expression: For Actual Pressure (P_3) and Temperature (T_3) Corrections	Expression: For Use when Geographic Average Barometric Pressure and Seasonal Average Temperature Have Been Incorporated into the Sampler Calibration
Mass flowmeter	I	I
Orifice and pressure indicator	$I\left(\frac{P_3}{P_{std}}\right)\left(\frac{298}{T_3}\right)$	I
Rotameter, or orifice and pressure recorder having a square root scale*	$I\sqrt{\left(\frac{P_3}{P_{std}}\right)\left(\frac{298}{T_3}\right)}$	I

*This scale is recognizable by its nonuniform divisions and is the most commonly available for high-volume samplers.

to a graph of the calibration curve for the sampler's flow rate measuring device or by calculating initial and final Q_{std} using Equation 3.1a. After the initial and final sampling standard flow rates have been determined, calculate the average sampling standard flow rate (Q_{std}) using the following equation:

$$Q_{std} = \frac{Q_I + Q_F}{2} \tag{3.1b}$$

where:

Q_{std} = standard volumetric flow rate, std m^3/min
Q_I = initial sampling standard flow rate, std m^3/min
Q_F = final sampling standard flow rate, std m^3/min

(22) Calculate the total air volume sampled using the following equation:

$$V = Q_{std} \times t \tag{3.2}$$

where:

V = total air volume sampled, std m^3
Q_{std} = average sampling standard flow rate, std m^3/min
t = sampling time, min

(23) Calculate and report the particulate matter concentration using the following equation:

$$TSP = \frac{(W_f - W_i)10^6}{V} \tag{3.3}$$

where:

TSP = mass concentration of total suspended particulate matter, μg/std m^3
W_f = final weight of exposed filter, g
W_i = initial weight of clean filter, g
V = air volume sampled, std m^3
10^6 = conversion of g to μg

(24) If desired, the actual particulate matter concentration can be calculated as follows:

$$(TSP)_a = TSP(P_3/P_{std})(298/T_3) \tag{3.4}$$

where:

$(TSP)_a$ = actual concentration at field conditions, μg/m^3
TSP = concentration at standard conditions, μg/std m^3
P_3 = average barometric pressure during sampling period, mm Hg
P_{std} = 760 mm Hg (or 101 kPa)
T_3 = average ambient temperature during sampling period, K

EXPERIMENT GENERAL

Total Suspended Particulate (TSP) is a criteria pollutant for which the United States Environmental Protection Agency (USEPA) has established standards to protect human health. The health standard is 75 μg/m^3 on an annual average basis, or 260 μg/m^3 for 24 hours, not to be exceeded more than once per year.

Air quality is measured by a high volume air sampler (hi-vol). This device draws ambient air through a preweighed fiberglass filter. As particulate matter builds on the filter, the air flow is restricted.

Equipment to use is a:

- GMW 2310 hi-vol sampler
- flowmeter (40–60 cfm) or equivalent

PROCEDURE (CONDENSED FROM ILLINOIS EPA STANDARD OPERATING PROCEDURE FO-053)

(1) Read 40 CFR, Part 50, Appendix B.

(2) Complete the hi-vol data form that is Table 3.3. Note the following:
- Collected by – full name of student
- Hi-vol motor no. – serial number of hi-vol motor installed
- Filter no. – serial number of filter from filter box
- Data – actual day the sample was collected (normally the same day for both start and stop times)
- Time – start and stop time of sample collection (Use 24-hour clock time. Normal sampling begins at 0000 hours on the sampling data and ends at 2400 hours on the same date.)

(3) Pre-Sampling Operations:
- Raise lid and secure to rear catch with the aluminum strip.
- Remove face plate by loosening the four wing nuts, allowing the swing bolts to swing down and out of the way.
- Wipe any dirt accumulation from around filter holder with a clean cloth.
- Wipe face plate gasket clean.
- Carefully center a new filter, rougher side up, on the supporting screen. Properly align the filter on the screen, so that when the face plate is in position, the gasket will form an airtight seal on the outer edges of the filter.
- Turn the motor on to hold the filter in place. Secure the filter with the face plate and four brass swing bolts with sufficient pressure to avoid air leakage at the edges. Diagonally opposite wing nuts should be tightened simultaneously. Do not over-tighten, because gasket could deform.
- Close lid carefully and secure catch with aluminum strip or padlock.
- Allow sampler motor to continue running for 5 minutes for warm up.
- Connect the flowmeter calibrated with the sampler to the brass pressure tap at the bottom of the motor/blower unit.
- Hold flowmeter at eye level in a vertical position and read the widest part of the float. Record flowmeter reading and conversion to CFM from the calibration table.

  Note: Flow measurements are taken at the beginning and end of the sampling period. Disconnect flowmeter during sampling period to prevent excessive clogging.
- Turn off motor and set timing mechanism to start at midnight of

TABLE 3.3. Hi-Vol Data Form.

HI-VOL DATA FORM Table 3.3

Environmental Protection Agency
Air Pollution Control

Important - Fill In All Field Information

City ____________
Building ____________
Street ____________
Collected by ____________
Hi-Vol Motor No. ____________
Flowmeter No. ____________
Filter No. ____________
Sampling Date ____________
Time Start ________ Stop ________

☐ Site inoperative or Malfunctioned

	Start	Stop
Flowmeter Reading	____	____
Flow Rate (CFM)	____	____

Elapsed Time Meter Start ☐☐☐☐☐☐

Stop ☐☐☐☐☐☐

FOR LABORATORY USE ONLY

Filter Weights Final ________

Initial ________

TSP ________ ug/m3

Lab ID. No. ____________
Date Sample Rec'd ____________
Date Analysis Checked ____________
Date TSP Data Forwarded ____________
Date Filter Forwarded ____________

If Voided, See Reverse Side

SAMPLE COLLECTION INSTRUCTIONS

1. Measure stop flow rate.*
2. Remove exposed filter.
3. Clean shelter and faceplate gasket.
4. Install clean filter
5. Measure start flow rate.*
6. Set timer for next scheduled sampling date.
7. Secure shelter.
8. Record all requested data on the hi-vol data form or line out if not applicable.
9. Mail filter and data form to laboratory immediately.
10. If you have any questions or need assistance, call collect to 217/782-5811.

* Should measured flow rates be below 40 CFM or above 60 CFM please contact the Illinois EPA as soon as possible.

Comments: ____________

VOIDED

☐ Greater than CM2 of filter missing
☐ Flow rate less than 35 or greater than 60 CFM
☐ Sampling time less than 23 hours or greater than 25 hours
☐ Poor gasket seating
☐ Filter number received does not correspond to data card
☐ Missing information ____________
☐ Other:

next scheduled sampling period. Record initial elapsed time reading.

(4) Post Sampling Operations

- Open the door to the timer. Turn on the motor and allow it to warm up for five minutes. While the motor is running, connect the flowmeter to the sampler with the tubing supplied. When stabilized, hold flowmeter in vertical position and read. Record the reading on Data Form in the Flowmeter Reading Stop column. Turn off the motor. Convert the reading using the conversion chart (refer to Table 3.4) and record it on the Flow Rate (CFM) Stop column of Table 3.3.
- Open shelter, loosen wing nuts, and remove face plate. Fold card lengthwise with printing on the outside. Slip lower edge of card

TABLE 3.4. **Hi-Vol Sampler Calibration Table.**

Site Location:			Hi-Vol I.D. No.: 16501
Site Address:			
Operator:			Flowmeter No.: 16501
Installation Date:			Calibration Date: 12/10/80
Flowmeter Reading	**Flow Rate (CFM @STP)**	**Flowmeter Reading**	**Flow Rate (CFM @STP)**
70	See footnote	51	50.9
69	See footnote	50	49.8
68	See footnote	49	48.8
67	See footnote	48	47.8
66	See footnote	47	46.7
65	See footnote	46	45.7
64	See footnote	45	44.6
63	See footnote	44	43.6
62	See footnote	43	42.6
61	See footnote	42	41.5
60	See footnote	41	40.5
59	59.2	40	See footnote
58	58.1	39	See footnote
57	57.1	38	See footnote
56	56.1	37	See footnote
55	55.0	36	See footnote
54	54.0	35	See footnote
53	52.9	34	See footnote
52	51.9	33	See footnote

*Corr. coeff.—.999, slope—.964, stnd. error—250, intercept—1.870.
The flow rate is out of range. Note this condition in comments portion of filter card and record flowmeter readings.

under exposed filter and raise it from support screen. Fold the filter lengthwise, so that sample touches sample and close the folded card over it for protection. Affix two small adhesive tabs over the edge to hold the card closed. These steps help guard against touching the filter or losing it in the wind. Insert in envelope.

- Record final elapsed time reading (should be 1440 ± 60 min).
- Record all unusual occurrences in the comments section.
- Reset sampler next sampling sequence as per third item under Pre-Sampling Operations.

DISCUSSION POINTS

What are factors that may induce errors in the sampling process that are addressed in the design of the hi-vol cabinet? What factors *could* be addressed in design improvements?

If the sampler is located in an area that has a high percentage of organic dust (i.e., sawmill, grain drying, etc.) would bacteria growth on the sample increase, decrease, or have no effect on sample weight? What steps could be taken to minimize the effects of bacterial growth?

What sizes of particles are likely *not* to be collected in the hi-vol sampling device? Why?

FLOW CALIBRATION OF A HIGH-VOLUME SAMPLER PARTICULATE COLLECTION DEVICE—EXPERIMENT

INTRODUCTION

Note: If you have not performed the previous experiment, read this section before starting. This experiment is to demonstrate how flow rate can be determined and used to calibrate an air pollution sampling device; the example used here is a high-volume particulate sampler (hi-vol).

The two key elements in determining the concentration of an air pollutant are: (1) an accurate measurement of the amount of pollutant which has been sampled, (2) and an accurate measurement of the gas media through which the pollutant has been transported. This experiment will utilize tools generally available in the laboratory to check the flow rate of a hi-vol particulate sampling device. Upon completion, the student should be familiar with the basic concepts involving volume flow of gases and the effect of flow rate on air pollution concentration levels.

Equipment necessary for this work is:

- hi-vol sampler with magnehelic gauge

- Reference Flow Device (RFD)
- calibrated oil manometer (± 0.05 inch H_2O)
- resistance plates
- thermometer ($\pm 0.05°C$)
- also, read 40 CFR, Part 50, Appendix B (See Appendix A of this manual.)

PRE-AUDIT PROCEDURES (CONDENSED FROM ILLINOIS EPA STANDARD OPERATING PROCEDURE QA-014)

(1) Open hi-vol shelter hood and secure it to the back latch.

(2) Attach the thermometer and manometer support braces to the top edge of the shelter, and attach the thermometer and manometer to the respective support braces.

(3) Open both ports on the manometer by turning the L-connectors 3/4 revolution counterclockwise, then connect a 2′ section of 3/16″ ID latex hose to one of the ports.

(4) Check the manometer liquid for free movement against pressure, and adjust the manometer scale to zero.

Note: To ensure a uniform method of measurement, the magnehelic gauge must be placed in an upright position.

(5) Check the meter zero on the magnehelic gauge and adjust if necessary.

(6) Connect a 2′ section of 3/16″ ID latex hose from the hi-vol motor pressure port to the magnehelic gauge inlet.

(7) Check for the presence of a glass fiber filter on the hi-vol's filter holder. If a filter is absent, then remove the hold down frame on the filter holder and proceed to Step 12. If the filter is unexposed, then remove the hold down frame on the filter holder and proceed to Step 13.

(8) Engage the hi-vol power switch, and allow the system to warm up for ten minutes.

(9) Record the magnehelic meter reading.

(10) Disengage the power switch, and record the elapsed time reading.

(11) Remove the exposed filter, fold it down the center, and carefully place it in an envelope for future analysis.

(12) Place an unexposed filter on the filter holder.

(13) Place the REF device over the filter, and secure it to the filter holder by tightening the lower side wing nuts by hand.

(14) Open the REF lid and place the single hole resistance plate in position.

(15) Engage the hi-vol power switch.

(16) Plug the resistance plate hole with a finger, and listen for air leaks. Do not exceed two minutes.

Note: If an air leak is suspected in the REF device or between the REF device and filter holder, try to correct the problem. If an air leak is suspected elsewhere, then continue with the audit, but record the observation on the audit form.

(17) Disengage the hi-vol power switch.

(18) Connect the free end of the 3/16" ID latex hose on the manometer to the REF manometer port.

(19) Remove the single hole resistance plate.

(20) Secure the REF lid by tightening the lid wing nuts by hand and position the wind deflector atop the lid.

AUDIT PROCEDURES

(1) Engage the hi-vol power switch, and allow at least ten minutes for the motor to warm up.

(2) Read and record the manometer displacement to within ±0.05 inch, the magnehelic meter to within ±0.1 inch, and the thermometer to within ±0.5°C.

(3) Disengage the hi-vol power switch.

POST-AUDIT PROCEDURES

(1) Disconnect the latex hose from the manometer and REF device, then close the manometer ports.

(2) Remove the REF device, wind deflector, manometer, thermometer, glass fiber filter, and thermometer and manometer support braces.

(3) Place an unexposed, preweighed glass fiber filter on the hi-vol's filter holder.

(4) Reattach the hold down frame to the filter holder, and firmly tighten the side wing nuts by hand.

(5) Engage the hi-vol power switch, and allow ten minutes for the system to warm up.

(6) Read and record the magnehelic meter to within ±0.1 inch.

(7) Disengage the hi-vol power switch, and record the elapsed timer reading.

(8) Complete the High Volume Data Form, Attachment 1.

(9) Remove the latex hose from the magnehelic gauge and hi-vol motor.
(10) Place the hose and magnehelic gauge in their original locations.
(11) Close and secure the hi-vol shelter hood.

CALCULATIONS

Calculations are described in Section 9.0 Calibration of 40 CFR 50, Appendix B. Following these instructions, prepare the calibration curve describing measured flow rate vs. actual flow rate, using the appropriate formula from Table 3.5. Use at least four data points to plot the least squares fit calibration line.

If sample weight data are available, calculate the pollutant concentration.

EXTRACTION PROCEDURES FOR LEAD, CADMIUM, ARSENIC SULFATES AND NITRATES FROM AIR FILTERS

FILTER PREPARATION PROCEDURES

Purpose

Air filters sent to and received from field operators are inspected and logged in. The T.S.P. (Total Suspended Particulates) and the air flows (M^3/day) are calculated. Filters and data sheets are then filed for future analysis.

Procedure

(1) Envelopes are opened at the top, and the manila folder containing the filter and data sheet is carefully removed.
(2) The filter is opened and inspected for visual defects. Examples of a defect that would void a sampled filter are:
   - cracks or tears in filter
   - holes in filter
   - missing pieces of filter
   - evidence of gasket leakage
   - uneven folding of filter
   - evidence of lost particulate matter (such as dust in the envelope)
(3) Data sheets are checked for completeness ensuring that all necessary fields have been filled in by the field operator. If flow of information has not been given, the sample will be voided.
(4) The Saroad Code (station number) and date are entered on the back-

TABLE 3.5. High Volume Accuracy Audit Form.

Attachment A

A. STATION PARAMETERS:

Site Name:________________ Site Address:________________ Saroad Number:__________

Station Pressure:________mmHg Station Temperature:________ ^oC = ________ ^oK Seasonal Correction Factor (cf):______

Auditor Name:________________ Audit Date:__________

B. ANALYZER PARAMETERS:

Hi-Vol Shelter IEPA No.:________ Hi-Vol Motor IEPA No.:________ Motor Calibration Date:________

Magnehelic Gauge No.:________ Magnehelic Calibration Date:________

Flow Controller (Absent / Present) IEPA No.:________ Sample Saver (Absent / Present) IEPA No.:________

Timer (Absent / Elapsed / Programmable) IEPA No.:________

C. ORIFICE CALIBRATION UNIT PARAMETERS:

OCU IEPA No.:________ OCU Calibration Date:________ Manometer Type: Water / Oil

Q actual = ________ x $(H \times Ta/Pb)^{.5}$ + ________

D. AUDIT PARAMETERS:

Hi-Vol Magnehelic Gauge Reading:________in. H_2O Magnehelic Gauge Flowrate (Y):________scfm

OCU Manometer Reading:________in. H_2O OCU Q standard (X):________scfm Percent Difference (d_1):________%

E. MISCELLANEOUS PARAMETERS:

Original Filter Condition: Absent / Exposed / Unexposed Original Filter No.:__________

New Filter No.:__________

side of the data sheets. Samples requiring Alpha-Beta radiation analysis are marked with a "B" on the upper right corner of the sheet.

Flow data is entered on wedding data sheets from the seasonal average list.

(5) After equilibrating for more than 20 hours, filters are weighed for gross weight as information from the data sheets is entered into the terminal. Information is entered in the following order:

- filter number
- date of sampling
- motor number
- Saroad Code of sampling site
- parameter (either SSI or TSP)
- sample condition (good or bad)
- flowmeter readings (before and after)
- gross weight

After data is entered into the computer, the computer will display calculated flow in cubic meters per day (M^3/day), the T.S.P. (Calculated Parameter Value) and the tare weight of the filters. This information is then recorded on the data sheet.

The data sheets and filters are then returned to the folders for filing.

Filters that need radiation analysis are taken to the radiation counting facility.

(6) File boxes are made up each month. Filters are filed by Saroad Code. PM10 (SSI) and T.S.P. filters are filed in separate boxes. (Wedding filters are filed behind the PM10's). After radiation filters return from the counting facility, they can be filed.

Data sheets are filed in decks made up each month. Again, PM10 and T.S.P. data sheets are filed in separate decks.

Tare Weighing Procedure

(1) Filters are removed from boxes and stacked on a clean surface the day before weighing to equilibrate.

(2) Filters are visually inspected over a light table or black surface for defects. Filters are rejected for the following defects.

- pinholes
- discolorations
- dense spots
- dark spots
- torn or frayed edges

Filters are not rejected for:

- thin spots

- fiber detachment
- manufacturing lines

unless there is more than one such defect on filter surface.

(3) Filters are weighed on stirrup-pan balance by carefully bending filter along its length into a crescent shape and inserting it between the pan arms. Care should be taken not to crease the filter.

(4) Every eighth filter is reweighed for quality assurance. If a filter that is reweighed is more than .0002 grams from the original weight, the previous eight filters are reweighed.

(5) After filters are tare weighed, they are stacked in the glass cabinet until they are mailed out.

EXTRACTION PROCEDURE FOR Pb, Cd AND As

The extraction procedure for Lead, Cadmium and Arsenic (Pb, Cd, As) analysis of air filters is described here. The reference for this procedure is the *Federal Register*, Vol. 43, No. 194, Appendix G, 10-5-78: "Reference Method for the Determination of Lead in Suspended Particulate Matter Collected from Ambient Air." The purpose is to transfer any Pb, Cd or As which is on an air filter to a nitric acid/hydrochloric acid solution. A 3/4" × 8" strip cut from a high-volume, glass fiber, air filter is required.

Reagents include acids suitable for trace metal analysis. Make a solution which is 2.6 M in HNO_3 and .9 M in HCl. It is convenient to make approximately two liters and place it in a glass bottle equipped with a re-pipettor.

Procedures

(1) Using a pizza cutter, plexiglass fixture and a cardboard template, cut a 3/4" × 8" strip from the air filter. This is equivalent to 1/12th of the air filter.

(2) Carefully fold the filter strip and place it in a clean, disposable plastic 50 ml centrifuge tube.

(3) Add 15.0 mls of the HNO_3/HCl acid mixture and cap the tubes.

(4) Place the rack of tubes in the sonic bath and sonicate for 30 minutes.

(5) Remove tubes from the bath and add D.I. water to bring volume to 50.0 mls.

(6) Replace caps and gently mix samples, set aside for 30 minutes.

(7) Filter samples through a .45 micron filter. While filtering, rinse out tubes with D.I. water.

(8) Return filtered sample to the tubes.

Note: Filtering should be done as soon as possible after a 30-minute wait.

Note: When returning filtrate to tubes for arsenic analysis, volume should be measured.

Note: Samples are now ready to be analyzed for Pb, Cd or As by ordinary methods used for the respective metals, i.e., Flame AA for Pb, Cd; Arsine generation for As.

Note: Pb and Cd are normally run from the same tube—arsenic from a separate tube.

Note: A reference strip is run with each set 4 μg As, 10 μg Cd, 100 μg Pb are spotted on a blank strip of air filters.

Calculations

As, Pb, Cd (μg/ml)(final vol.)(12)/cubic meters air = μg/cubic meter.

SULFATES IN AIR FILTERS

Sulfate is extracted from glass fiber filter strips into deionized water. The sample is then filtered. The sample is passed through a cation-exchange column to remove interferences. The sample is then reacted with barium chloride at a pH of 2.5–3.0 to form barium sulfate. Excess barium reacts with methylthymol blue to form a blue-colored chelate at a pH of 12.5–13.0. The uncomplexed methylthymol blue color is gray; if it is all chelated with barium, the color is blue. The amount of uncomplexed methylthymol blue, measured at 460 nm is equal to the sulfate present. The reference for this work is "Sulfate in Water and Wastewater," Technicon Industrial Method No. 118-71WIB; revised January 1977.

One-twelfth of a glass pad (3/4 × 8 inch strip) is extracted into 50 ml deionized water. This amount of sample yields a detection limit of 2 mg/l $SO_4^{=}$.

Reagents and Equipment

(1) Nitrate/sulfate stock standard (800 μg ml NO_3^- and 100 μg/ml $SO_4^{=}$): Weigh 1.3046 g potassium nitrate (KNO_3) and 1.4797 g sodium sulfate (Na_2SO_4), which has been dried at 104°C for one hour and stored in a desiccator. Place in 1000 ml volumetric flask and dilute to one liter with deionized water. Add one ml of chloroform as a preservative. Store under refrigeration.

(2) Deionized degassed water: Fill a 4-liter vacuum flask to 3/4 capacity and stopper the flask. Draw a vacuum and stir the water with a magnetic stir plate for at least 10 minutes. This degassed water is used

to prepare all reagents for this method. It is also used as the wash water for the automated analysis.

(3) Barium chloride solution: Dissolve 1.526 g of barium chloride dihydrate ($BaCl_2 \cdot 2H_2O$) in about 600 ml degassed deionized water. Dilute to one liter and mix. Store in a plastic container.

(4) Hydrochloric acid, 1.0 *N*: Add 8.25 ml of concentrated hydrochloric acid (HCl) to 60 ml degassed deionized water and mix. Dilute to 100 ml and mix thoroughly. Store in a well-stoppered container.

(5) Alcohol, reagent grade, anhydrous

(6) Methylthymol blue color reagent (MTB): Transfer 0.1182 g to a 500 ml volumetric flask. Pipette 25.0 ml barium chloride solution into the flask and swirl to dissolve. Add 4.0 ml hydrochloric acid solution, 1.0 *N*, and swirl. Add 71 ml degassed deionized water and dilute to 500 ml with alcohol. Mix thoroughly. This reagent must be prepared fresh daily. It is recommended that the reagent be prepared at the end of the day previous to the day it will be used and allowed to stand overnight in a refrigerator.

(7) Sodium hydroxide, 0.18 *N*: Add 7.2 g sodium hydroxide (NaOH) to about 600 ml degassed deionized water and dilute to one liter. Mix thoroughly. Store in plastic container. When in use in the system, the opening of the bottle should be shielded (e.g., parafilm) to minimize the absorption of carbon dioxide from the air.

(8) Alkaline EDTA solution (for cleaning system): Dissolve 6.75 g of ammonium chloride (NH_4Cl) in about 600 ml deionized water. Add 57 ml of ammonium hydroxide (NH_4OH) and mix. Add 40 g EDTA, tetrasodium salt, and mix until dissolved. Dilute to one liter. Store in a plastic container.

(9) Cation exchange resin, Bio-Rex 70, 20-50 mesh, sodium form. Rinse resin with deionized water to remove fine particles. Store washed resin in stoppered bottle. Make sure resin is under deionized water.

(10) Technicon Autoanalyzed II with sulfate manifold, 30 6/l cam, 2 filters 460 nm

(11) Vacuum filtering apparatus with 0.45 m filters

(12) Ultrasonic Bath: Manufacturing by ALCAR Industries, Inc. 117 VAC, 60 Hz, cleaning power of 400 watts, cleaning intensity of 2 watts/in^2

(13) Plastic disposable centrifuge tubes, 50 ml

(14) Plastic template guide

(15) Fractional template with a line drawn parallel to the center line at 3/4 inch distance

(16) Pizza cutter: Thin wheel, thickness <1 nm

Sample Preparation Procedure

(1) Cut a 3/4 × 8 inch strip from the air filter using the template, pizza cutter and plastic template guide.
(2) Place the filter in a plastic centrifuge tube and fill to 50 ml with deionized water. The eighth position in the rack should contain a blank filter strip, the 17th position a duplicate and the 24th a spike. (Spike amount = 0.5 ml of 1000 μg/ml $SO_4^{=}$.)
(3) Adjust the water volume in the ultrasonic bath to match the water level in the centrifuge tubes.
(4) Subject the samples to ultrasonic vibration for 45 minutes.
(5) Remove the samples from the ultrasonic bath and cool for 30 minutes.
(6) Filter the samples before analyzing on the Technicon autoanalyzer.

Preparation of Ion-Exchange Column

(1) Insert a small plug of glass wool into a 10-inch length of 0.081 I.D. Tygon tubing.
(2) Fill the tubing with distilled water and transfer the resin to the column with a disposable plastic eyedropper.
(3) A second plug of glass wool is inserted into the column.
(4) Connect the column to the manifold while the pump is running. The column should be replaced daily, or if air enters the system.

Linearizing the Calibration Curve

(1) The amount of methylthymol blue used in the color reagent must be determined experimentally for each lot of methylthymol blue purchased. Because of the low purity of the commercial dye, barium to MTB ratios of 0.9:1 or lower are usually needed for obtaining linear calibration curves.
(2) Prepare barium chloride solution by dissolving 1.526 g $BaCl_2 \cdot 2H_2O$ in one liter of degassed deionized water.
(3) Dissolve 0.4477 g MTB in 100 ml degassed, deionized water.
(4) While keeping the amount of barium constant, varying amounts of the MTB solution are added to 250 ml volumetric flasks. Two ml of 1.0 *N* HCl and enough degassed deionized water is added to bring the volume to 50 ml. The solutions are diluted to 250 ml with ethanol. Refrigerate overnight before running on the technicon.

ml Ba Soln	ml MTB	ml HCl	ml D.I.	ml Ethanol
12.5	10	2.0	25.5	200
12.5	11	2.0	24.5	200
12.5	12.5	2.0	23.0	200
12.5	13	2.0	22.5	200
12.5	14	2.0	21.5	200
12.5	15	2.0	20.5	200
12.5	16	2.0	19.5	200

(5) Run a standard curve and reference of the trail color reagents. The standard curve range is 2–40 mg/l $SO_4^=$.

(6) Calculate the correlation coefficient and % recovery for the reference for each color reagent used.

(7) Using the correlation coefficient and % recovery of the reference as criteria, an optimum amount of methylthymol blue can be selected for the preparation of the MTB reagent suitable for routine use.

$$\frac{.4774 \text{ g MTB} \times \text{Volume MTB in optimum trial color reagent}}{100 \text{ ml}}$$

$$= \text{g MTB for 250 ml}$$

$$\text{g MTB for 250 ml} \times 2 = \text{g for routine batch of 500 ml}$$

Standard Curve Preparation

Prepare standards in 100 ml volumetric flasks. Dilute to volume with deionized water. Standards contain no preservative and should be refrigerated if they are to be used the next day. Standards should be discarded after 48 hours.

ml 1000 μg/ml SO_4	Final Volume	Concentration
0.2	100	2.0 μg/ml
0.5	100	5.0 μg/ml
1.0	100	10.0 μg/ml
2.0	100	20.0 μg/ml
3.0	100	30.0 μg/ml
3.5	100	35.0 μg/ml
4.0	100	40.0 μg/ml

Analysis

(1) Place all lines in their reagent containers and pump reagents through the manifold for 15 minutes.
(2) Attach the ion exchange column to the manifold. It is important that no air bubbles enter the ion-exchange column at any time. If air bubbles become trapped, it is advisable that a new column be prepared.
(3) Set baseline with zero control or aperture.
(4) Use standard cal to set 50% full scale when midpoint standard is in light path.
(5) A standard curve must be run for each tray of 40 cups. Recheck reagent zero and midpoint standard after 20 cups.
(6) At the end of the day, the system should be washed with a solution of EDTA. Place the MTB and sodium hydroxide line in water for a few minutes and then into the tetrasodium EDTA for 10 minutes. Wash the system with water for 15 minutes before shutting down. Pump manifold dry if it is not to be used for several weeks.

Calculations

Run linear regression on calculator using mg/l vs. peak height. Compute correlation coefficient, slope and linear regression points.

Spike Calculation:

$$\%\ \text{recovery} = \frac{\text{mg/l spike} \times 50.5\ \text{ml}}{\text{mg/l sample} \times 50\ \text{ml} + 500\ \mu\text{g}}$$

Duplicate Calculation:

$$\%\ \text{difference} = \frac{\text{duplicate} - \text{original} \times 100}{\text{average}}$$

Clean-Up Requirements

(1) Centrifuge tubes and cups are to be emptied and thrown in the trash.
(2) Discard unused MTB color reagent. This reagent must be made fresh the night before you are to run sulfates.

Miscellaneous Notes and Comments

(1) Do not use a surfactant in any of the reagents.
(2) All water utilized in this system should be degassed prior to use.

(3) If baseline is noisy, degass reagents or discard reagents and make again with degassed water.

(4) The red/red pump tube for the MTB color reagent *must be made of silicone tubing*. The ethanol in the color reagent attacks the walls of the usual type pump tubing, and the amount of color reagent does not remain constant.

NITRATES IN AIR FILTERS

Nitrate is extracted from glass fiber filter strips into deionized water. The sample is then filtered. The sample is passed through a column containing copper-coated cadmium to reduce the nitrate to nitrite. The nitrite ion then reacts with sulfanilamide under acidic conditions to form a diazo compound. This compound then couples with N-1-naphthylethylenediamine dihydrochloride to form a reddish purple azo dye, that is measured colorimetrically at 520 nm. The reference for this work is "Nitrate and Nitrite in Water and Wastewater," Technicon Industrial Method NO 100-70 W/B, revised: January 1978.

One-twelfth of a glass pad (3/4 × 8 inch strip) is extracted into 50 ml deionized water. This amount of sample yields a detection limit of 1.6 mg/l NO_3^-.

Reagents and Equipment

(1) Nitrate/sulfate stock standard (800 μg/ml NO_3^- and 1000 μg/ml $SO_4^{=}$): Weigh 1.3046 g potassium nitrate (KNO_3) and 1.4797 g sodium sulfate (Na_2SO_4), which has been dried at 104°C for one hour and stored in a dessicator. Place in 1000 ml volumetric flask and dilute to one liter with deionized water. Add one ml of chloroform as a preservative. Store under refrigeration.

(2) Brig wash solution: 1 ml Brig-35 per 2 liters of deionized water
Note: More recent procedure is to simply use deionized water.

(3) Ammonium chloride buffer: Dissolve 20 g of ammonium chloride (NH_4Cl) and 0.2 g disodium EDTA in 2 liters of deionized water. Adjust pH to 8.5 with concentrated ammonium hydroxide (NH_4OH). Add 1 ml of Brig-35 per 2 liters.

(4) Dilution water: Add 1 ml of Brig-35 per 2 liters of deionized water.

(5) Color reagent: To 750 ml deionized water add 100 ml concentrated phosphoric acid (H_3PO_4) and 10 g of sulfanilamide. Add 0.5 g of *N*-1-naphthylethylenediamine dihydrochloride. Stir to dissolve. Dilute to one liter with deionized water. Add 0.5 ml Brig-35. Store in dark bottle and refrigerate.

(6) Technicon Autoanalyzer II with nitrate manifold, 40 4/1 cam, 2 filter 520 nm.
(7) Vacuum filtering apparatus with 0.45 μm filters.
(8) Ultrasonic bath: Manufactured by ALCAR Industries, Inc., 117 VAC, 60 Hz, cleaning power of 400 watts, cleaning intensity of 2 watts/in^2
(9) Plastic disposable centrifuge tubes, 50 ml.
(10) Plastic template guide
(11) Fractional template with a line drawn parallel to the center line at 3/4 inch distance
(12) Pizza cutter: Thin wheel, thickness <1 mm
(13) Cadmium metal. Particles are used in the column which pass a 25 mesh sieve, but are held back by a 60 mesh sieve.
(14) 6 *N* hydrochloric acid. Add 500 ml concentrated hydrochloric acid (HCl) to 500 ml deionized water.
(15) Copper sulfate, 2% W/V. Dissolve 20 g copper sulfate ($CuSO_4 \cdot 5H_2O$) in one liter of deionized water.

Sample Preparation Procedure

(1) Cut a 3/4 × 8 inch strip from the air filter using the template, pizza cutter and plastic template guide.
(2) Place the filter in a plastic centrifuge tube and fill to 50 ml with deionized water. The eighth position in the rack should contain a blank filter strip, the 16th position, a duplicate and the 24th a spike. (Spike amount = 0.5 ml of 800 μg/ml.)
(3) Adjust the water volume in the ultrasonic tank such that the tank water level matches the water level of the centrifuge tubes.
(4) Subject the samples to ultrasonic vibration for 45 minutes.
(5) Remove the samples from the ultrasonic bath and cool for 30 minutes.
(6) Filter the samples before analyzing on the Technicon auto analyzer.

Preparation of Cadmium Reductor Column

(1) Cadmium metal is sized. Particles are used in the column which pass a 25 mesh sieve, but are held back by a 60 mesh sieve.
(2) Approximately 10 g of cadmium is cleaned with 50 ml 6 *N* HCl. Decant the HCl and wash with another 50 ml HCl. Decant and rinse with deionized water. Cadmium particles should now be a silver color.
(3) Swirl cadmium in 50 ml 2% copper sulfate until blue color fades. Decant and repeat until a brown colloidal precipitate forms.

(4) Wash the copper-coated cadmium with deionized water. The copper-coated cadmium should be black.

(5) The reduction column is a 14-inch length of 0.090 I.D. Tygon tubing. Plug one end with a small amount of glass wool. Fill the tubing with deionized water to prevent entrapment of air bubbles during packing the column. Transfer the copper-coated cadmium to a funnel attached to the column with a small spatula. When the column is filled, place another small glass wool plug in the end. Be careful not to allow any air bubbles to be trapped in the column.

Analysis

(1) Standard curve preparation
- Prepare standards in 100 ml volumetric flasks. Dilute to volume with deionized water. Standards contain no preservative and should be refrigerated if they are to be used the next day. Standards should be discarded after 48 hours.

ml 800 μg/ml $NO_3^=$	Final Volume	Concentration
0.2	100 ml	1.6 μg/ml
0.5	100 ml	4.0 μg/ml
1.0	100 ml	8.0 μg/ml
2.0	100 ml	16.0 μg/ml
3.0	100 ml	24.0 μg/ml
3.5	100 ml	28.0 μg/ml
4.0	100 ml	32.0 μg/ml

(2) Start up
- Pump Brig-wash, dilution water and buffer through the manifold for 15 minutes.
- Attach cadmium column in manifold. Condition the column with 100mg/l nitrate for 10 minutes, followed by 100 mg/l nitrite for 20 minutes.
- Sample wash for 5 minutes before placing color reagent line in color reagent container.
- Set baseline with zero control or aperture.
- Use standard cal to set 50% full scale when midpoint standard is in lightpath. Run zero and midpoint standard after 20 cups.
- A standard curve and recheck of zero and midpoint standard must be run for each tray of 40 cups.

(3) Shut down

- Detach cadmium column and close together with plastic nipple. Store in 250 ml bottle of buffer.
- Place all lines in wash solution and wash out system for 15 minutes.
- Pump system dry if it is not going to be used again for a few weeks.

Calculations

Run linear regression on calculator using mg/l vs peak height. Compute correlation coefficient, slope and linear regression points.

Spike calculation:

$$\% \text{ recovery} = \frac{\text{mg/spike} \times 50.5 \text{ ml}}{\text{mg/l sample} \times 50 \text{ ml} + 400\ \mu\text{g}} \times 100$$

Clean-Up Requirements

Cups and centrifuge tubes are emptied and are thrown in the trash.

Miscellaneous Notes and Comments

(1) Do not add deionized water to filters in centrifuge tubes until just before you are ready to place them in the ultrasonic bath.

(2) Filter samples 30 minutes after being taken out of the ultrasonic bath. Do not allow samples to sit unfiltered for more than a few hours at most.

(3) Extracted samples are neutral and must be run within 48 hours of extracting. Refrigerate samples overnight if they are not run the same day that they are extracted.

LEAD ANALYSIS—EXPERIMENT

Introduction

Lead enters the biosphere from lead-bearing minerals in the lithosphere through both natural and human-mediated processes. In natural processes, lead is first incorporated in soil in the active soil zone, from which it may be absorbed by plants, leached into surface waters, or eroded into windborne dusts. In addition, minute amounts of radioactive lead reach the atmosphere through the decay of radon gas released from the earth.

The lead used in gasoline antiknock additives represents the major human-mediated sources of lead emissions to the atmosphere. Additional important sources of lead emissions include stationary combustion of waste oil and incineration of solid waste. Other stationary sources include primary and secondary lead smelting, battery manufacturing and lead alkyl manufacturing.

Lead is emitted to the atmosphere primarily in the form of inorganic particulates. However, small amounts of organic vapors have been reported in the vicinity of gasoline service stations, garages and heavy traffic areas. These organic vapors undergo photochemical decomposition in the atmosphere, but they may also be adsorbed on dust particle surfaces.

Principle of Measurement

Suspended particulate matter from the ambient air is collected on a glass-fiber filter for 24 hours using a high-volume air sampler. Lead in the particulate matter on the filter is extracted with nitric acid (HNO_3), facilitated by heat or by a mixture of HNO_3 and hydrochloric acid (HCl) with ultrasonication. The lead content of the sampled is analyzed by atomic absorption spectrometry using an air-acetylene flame at the 283.3 or 217.0 nm lead absorption line.

Calibration of the atomic absorption spectrophotometer is carried out with lead nitrite [$Pb(NO_3)_2$] in standard solutions of different concentrations.

Atomic Absorption Spectrometry (AAS)

In an atomic absorption spectrophotometer, a portion of the sample is aspirated into a flame causing chemical reduction of metal ions that produces ground state atoms. The ground state atoms, then, absorb a portion of the light that is passed through the flame and the absorbance is measured. In a sense, we may speak of the flame as parallel to the absorption cell used in other types of spectrophotometers (see Figure 3.4).

Atomic absorption spectra are line spectra, unlike molecular spectra which are in the form of broad bands. The atomic absorptions are discrete lines of relatively narrow bandwidth at wavelengths which are characteristic of the given element.

The source of light is a hollow cathode lamp which contains a cathode coated with the metal to be measured. This light source limits light at the discrete absorption wavelengths. Introduction of sample into the flame causes absorption of some of the incident light. After passing through the monochromator, which selects one absorption line, the reduction in beam intensity is measured by the detector.

In double beam spectrophotometers, a second reference light beam from the same hollow cathode lamp bypasses the flame and is received at the

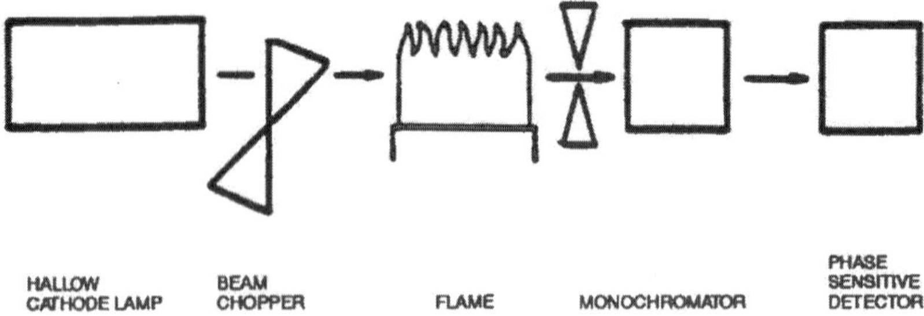

Figure 3.4 A simple atomic absorption spectrometer.

detector, in alternate pulses to the sample beam. The ratio of beam intensities, after passage through the monochromator, is measured.

The absorption laws applying to spectrophotometric measurements also apply to atomic absorption. Beer's law (the linear relationship between absorbance and sample concentration) applies to most measurements using this technique, but deviations occur, particularly at higher concentrations. Thus, it is normally advisable to calibrate over the full range of expected concentrations. Additionally, since instrumental parameters are liable to change, it is normal practice to calibrate the instrument frequently.

Objectives and Procedure

In this experiment, you will extract lead from a hi-vol filter and measure the concentration with an atomic absorption spectrophotometer.

The procedure for the measurement of lead in ambient suspended particulate is detailed in Title 40, Part 50 (Appendix G) of the U.S. Code of Federal Regulations.

References

- CFR 40, Part 50, Appendix G
- Roger Perry and Robert S. Young, *Handbook of Air Pollution Analysis*
- Air Quality Criteria for Lead, EPA-600/8-77-017, December 1977

CONTINUOUS CARBON MONOXIDE ANALYZER—EXPERIMENT

BACKGROUND

The purpose of this experiment is to familiarize the researcher with the operation of a gaseous pollutant monitoring device, its principles of operation and factors that affect the quality of data obtained from such instruments. A Bendix infrared Gas Analyzer will be used in the example.

Some gases are opaque under normal lighting conditions. The most common example is nitrogen dioxide. Although most gases appear to be transparent, however, many gases, such as carbon monoxide, attenuate (i.e., absorb) electromagnetic radiation at wavelengths outside of the visible light range. Carbon monoxide (CO) absorbs infrared light energy; its ability to absorb infrared allows its detection with a device that emits infrared light and a photo cell that can detect small changes in infrared energy that is transmitted through the gas sample.

Equipment

- Bendix model 8501-5CA infrared analyzer
- instruction manual
- calibration gases (0, 9 and 43 ppm CO in air with 350 ppm CO_2)
- CO gas regulator (CGA 350)
- bypass flowmeter (0 to 1.0 liter/min)
- digital voltmeter (DVM)

  Note: All analyzer response readings are to be taken from a digital voltmeter and strip chart recorder attached to the output terminals of the analyzer.

References

- *Federal Register*, Vol. 47, No. 234, Appendix C, 12-8-82: "Measurement Principle and Calibration Procedure for the Measurement of Carbon Monoxide in the Atmosphere (Non-Dispersive Infrared Photometry)"

PROCEDURES (CONDENSED FROM ILLINOIS EPA PROCEDURES FO-002 AND QA-050)

Pre-Audit

(1) Review 40 CFR 50, Appendix C and 40 CFR 58, Appendix A.

(2) Secure each gas cylinder to a firm support brace in close proximity to the CO analyzer.

(3) Attach voltmeter and strip chart to analyzer output leads.

Perform Zero Check

(1) Open cylinder valve on zero air gas cylinder.

(2) Open output valve on zero air regulator.

(3) Place three-position switch on "ZERO" position.

(4) Adjust regulator output valve to achieve a flow rate through the analyzer as indicated on the shipping certificate (usually a reading between 3 and 5 on the flowmeter).

(5) Observe the strip chart recorder and DVM readings until stable.

(6) Record readings on the Maintenance Performance Verification (MPV) log and on the strip chart.

(7) Return three-position switch to "SAMPLE" position.

(8) Close output valve on gas cylinder regulator.

(9) Close cylinder valve.

Note: Closing valves in the sequence allows the regulator to remain pressurized with cylinder gas.

Perform Span Checks

(1) After ensuring that the pressure regulator valve and output port valve are still closed, attach the regulator to one of the audit cylinders.

(2) Open the gas cylinder valve and leak check the regulator/gas cylinder connection with a soap solution. If a leak is detected, tighten the regulator connecting nut until the leak has stopped.

(3) Observe the regulator high pressure gauge reading.

Note: Record all audit data on the CO Accuracy Audit Form, Table 3.6.

(4) Attach a Drierite cartridge to the station sample line for 30 minutes.

(5) Connect a combination of Teflon lines to the station sample line, tee, rotameter and regulator as shown in Figure 3.5.

(6) Adjust the pressure regulator output valve until the low pressure gauge reads 50 psi.

(7) Slowly open the regulator output port valve until a flow rate of approximately 1.0 l/m is indicated on the excess flow rotameter.

Note: Adjust the regulator output port valve as needed to maintain an excess flow near 1.0 l/m.

(8) Monitor the recorder output until stable, then record the corresponding gas cylinder number and cylinder CO concentration.

(9) Close the gas cylinder valve, and allow the high and low pressure gauges to fall to zero psi.

(10) Close the pressure regulator and output port valves.

TABLE 3.6. CO Accuracy Audit Form No. 1.

AUDITOR: ______________ DATE: ______________ ROOM TEMP.: ______________

SITE: ______________ SAROAD NO.: ______________

ANALYZER MAKE & MODEL: ______________ IEPA NO.: ______________

RANGE: ______________ ANALYZER STATUS: ______________

TELEMETRY SLOPE:		TELEMETRY INTERCEPT:		
CYLINDER NUMBER	CYLINDER CO CONCENTRATION (ppm)	RAW SCAN (ppm)	CORRECTED RAW SCAN (ppm)	PERCENT DIFFERENCE

(11) Repeat Steps 1 – 3 and 5 – 10 for each audit cylinder.

Post-Audit

(1) Disassemble the Teflon audit lines.
(2) Remove the pressure regulator from the audit cylinder.
(3) Cap each audit cylinder port and both ports on the CO regulator.
(4) Reconnect the station CO sample line to the station manifold. Allow 30 minutes for ambient sampling.

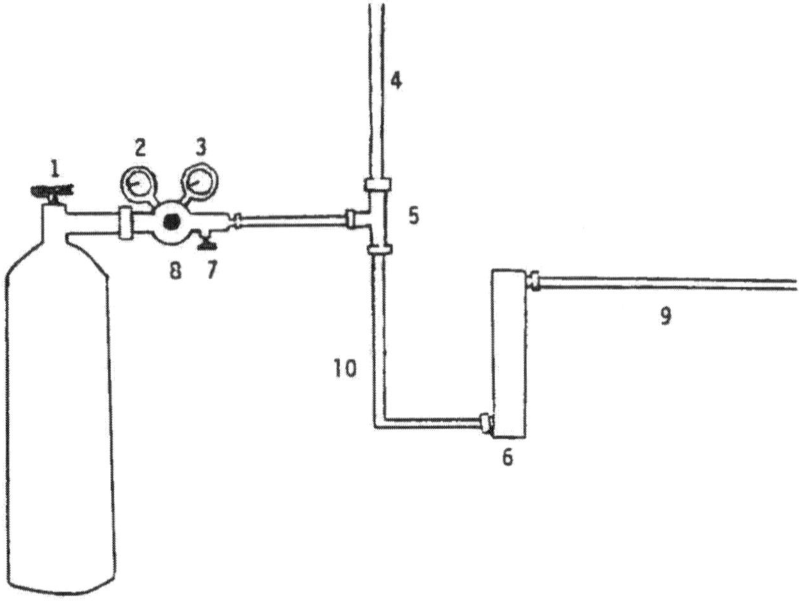

1. Gas cylinder valve
2. High pressure gauge
3. Low pressure gauge
4. CO station sample line
5. Teflon tee
6. Rotameter
7. Regulator output port valve
8. Pressure regulator valve
9. Teflon vent line to outside
10. Teflon line

Figure 3.5.

Calculations

(1) Use a least squares fit between the CO reading from the instrument (the dependent variable), and the CO of the gas bottle (the independent variable) to solve for the coefficients of the following equation for a straight line.

$$Y = b_0 + mx$$

Discussion

Most instruments display some "nonlinearity." This is a characteristic which means that the calibration of the instrument against a known source does not follow a straight line. To what extent does the "curve" developed in this experiment not follow a straight line? Under what conditions is nonlinearity acceptable?

What is the accuracy of the calibration gas? What can be said of the overall accuracy of any readings that may be obtained from the instrument as you have calibrated it?

CONTINUOUS NO_2 MONITOR CALIBRATION—EXPERIMENT

INTRODUCTION

Calibration of atmospheric sampling equipment is necessary to ensure that the data generated by air monitors represent the actual concentration of pollutants in the air. The generation of standard test atmospheres (known contaminant concentrations) is essential to the calibration procedures for continuous air monitoring instrumentation. Calibration permits the development of a relationship between the concentration of gases to the monitor and the output instrument response.

Static and dynamic test atmospheres can be prepared by a variety of procedures. In the laboratory, you will prepare and use the following test atmospheres.

(1) A dynamic test atmosphere of cylinder NO using a dilution system.
(2) A dynamic test atmosphere of NO_2 using a permeation system in combination with a dilution system.
(3) A static test atmosphere of NO_2 in a nonrigid chamber.

Objectives are to:

(1) Prepare calibration curves of NO concentration vs instrument responses using NO cylinder gas/dilution dynamic test atmosphere.
(2) Use NO_2 permeation/dilution system to determine the efficiency of the NO_2-NO converter of the analyzer.
(3) Prepare and test a static test atmosphere of NO_2 in a nonrigid chamber.

EXPERIMENT 1—PREPARATION OF NO CALIBRATION CURVE

In this experiment, you will prepare various dynamic test atmospheres of nitric oxide (NO) with a cylinder of compressed NO and a dilution system. Calibration systems vary in complexity, but are usually variations of the system shown in Figure 3.6. Record data on Table 3.7.

More accurate calibrations can be done with the system in Figure 3.6 by

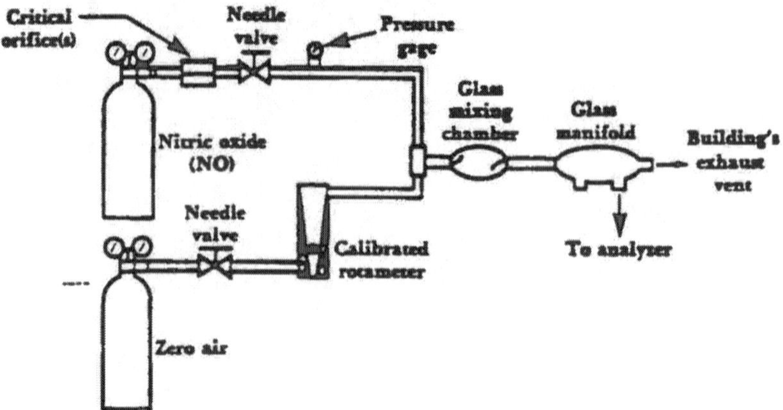

Figure 3.6 Schematic of a simple single dilution system.

replacing the rotameter with a more accurate flow measuring device (e.g., a combination of different range critical orifices). The NO channel of an NO-NO_x chemiluminescence analyzer, then, can be calibrated with the generated test atmospheres.

The concentration of NO in the test atmosphere is calculated by the following formula:

$$c_1 Q_s = c_2(Q_s + Q_z) \tag{3.5}$$

where:

c_1 = concentration of NO in standard NO cylinder, ppm
Q_s = flow from NO cylinder, sm^3/min
c_2 = concentration of NO in test atmosphere, ppm
Q_z = zero (diluent) air flow rate, sm^3/min

Since c_1 and Q_s have constant values in this experiment, the concentration of the test atmosphere (c_2) is changed by varying the flow of zero air (Q_z).

Procedures to follow are:

(1) Locate the calibration curves and necessary information (provided by the instructor) for the various flow measuring devices to be used.
(2) Read the Appendix on operation of the NO-NO_x analyzer.
(3) Make sure that the NO-NO_x analyzer is in the NO mode and on the 0–10 ppm range.
(4) Attach the analyzer inlet to the outlet of the calibration system.
(5) Switch and adjust the valve to provide zero air at the outlet. The total

air flow must exceed the total demand of the analyzer to ensure that no ambient air is pulled into the vent.

(6) Allow the analyzer to sample zero air until stable NO and NO_x responses are obtained.

(7) After the responses have stabilized, adjust the analyzer zero control.

(8) Adjust the NO flow from the standard NO cylinder to generate an NO concentration of approximately 80% of the upper range limit of the NO range. The exact NO concentration is calculated from Equation 3.5.

(9) Sample this NO concentration until the NO response has stabilized.

(10) Record c_2, Q_s, Q_z and net instrument response.

TABLE 3.7. Nitrogen Dioxide (Instrumental Lab) Calibration Data Form.

Name ____________________ **Analyzer serial no.** ____________________
Group no. ____________________ **Model no.** ____________________
Date ____________________

NO/NO_x Calibration and Linearity Check

Calibration points NO/NO_x	1 $F_D + F_O$, cm³/min	2 F_{NO}, cm³/min	3 $[NO]_{OUT}$, ppm	4 NO recorder, % scale	5 $[NO_x]_{OUT}$, ppm	6 NO_x recorder, % scale
Zero						
80% URL						
1						
2						

NO_2 Calibration by GPT

Calibration points NO_2	7 $[NO_2]$, ppm	8 $[NO]_{ORIG}$, ppm	9 $[NO]_{REM}$, ppm	10 $[NO_2]_{IMP}$, ppm	11 $[NO_2]_{OUT}$, ppm	12 NO_2 recorder, % scale
Zero						
ORIG						
80% URL						
1						
2						

(11) Repeat Steps 8 – 10 for other NO concentrations in order to construct a calibration curve by plotting concentration of NO on the x-axis and net instrument response on the y-axis of linear graph paper.
(12) Discuss the results, and derive Equation 3.5.

EXPERIMENT 2—DETERMINATION OF THE CONVERTER EFFICIENCY

The use of permeation techniques for preparation of standard mixtures is very useful for some contaminants. A permeation cell or tube consists of a gas confined above its liquid at a constant temperature. Under these circumstances, the gas will have a specific pressure and will permeate through certain materials at a constant rate, according to the concentration gradient that exists. Thus, a liquefied gas or volatile liquid sealed in a section of Teflon tubing, for example, when placed in a metered air stream can be used as a dynamic calibration standard. By passing different flow rates of diluent gas over the tube, gases of varying known concentration can be generated (see Figure 3.7).

The actual concentration of a sample gas can be calculated by the following formula:

$$c = \frac{\{PR\}\left\{\dfrac{24.46\ \mu\text{l}/\mu\text{-mole}}{M\ \mu\text{g}/\mu\text{-mole}}\right\}\left\{\dfrac{T\ °\text{K}}{298°\text{K}}\right\}\left\{\dfrac{760\ \text{mm Hg}}{P\ \text{mm Hg}}\right\}}{Q_T\ \text{l/min}} \tag{3.6}$$

where:

c = concentration of the sample gas, ppm
T = temperature of the system, °K
P = gas pressure of the system, mm Hg
PR = permeation rate, μg/min

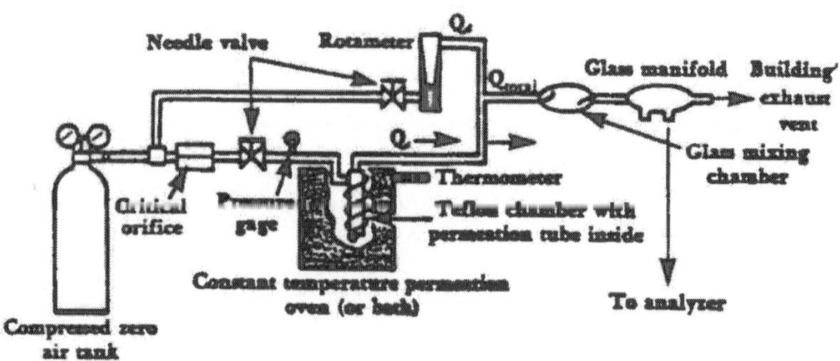

Figure 3.7 Permeation system.

Q_T = total flow rate, l/min
M = molecular weight of the permeation gas, μg/μ-mole
24.46 = molar volume of any gas at 25°C 760 mm Hg, μl/μ-mole

In this experiment, you will generate different concentrations of NO_2 from the permeation/dilution system in order to determine the NO_x responses of the analyzer.

Procedure to follow is:

(1) Record the room temperature and barometric pressure.

(2) Record the temperature of the permeation bath, and determine the permeation rate (PR) at this temperature from the calibration data provided by the instructor.

(3) Set the NO-NO_x analyzer in the NO_x mode on the 0 – 10 ppm range.

(4) Attach the analyzer inlet to the outlet of the calibration system.

(5) Set the zero air flow rate passing over the NO_2 permeation tube (Q_s) by adjusting the needle valve until the pressure differential across the orifice of 2 in. Hg is obtained. The orifice has been previously calibrated by your instructor. Use the calibration flow rate in your calculation.

(6) Calculate the approximate total flow rate (Q_T) needed to generate a test atmosphere with an approximate NO_2 concentration of 2 ppm using Equation 2, but neglecting temperature and pressure corrections. Thus,

$$Q_T = \frac{(PR)(24.46)}{(M)(c)} \tag{3.7}$$

where:

Q_T = required total flow rate, l/min
PR = NO_2 permeation rate, μg/min
M = molecular weight of NO_2 = 46 μg/μ-mole
c = desired NO_2 concentration = 2 ppm

(7) Determine the approximate required dilution flow rate (Q_z) in l/min.

$$Q_z = Q_T - Q_s$$

(8) Adjust the dilution air needle valve until the calculated dilution flow rate (Q_z) is obtained. It is not necessary to set this flow exactly. Determine the actual dilution flow rate (Q_z) from the rotameter reading.

(9) Record Q_t and Q_T.

(10) Calculate and record the actual concentration of generated NO_2 atmosphere using Equation 3.6, taking into account the temperature and pressure of the system.

(11) Sample this NO_2 concentration until the NO_x response has stabilized.

(12) Record net instrument response.

(13) Repeat Steps 6–12 for NO_2 concentrations of 4, 6 and 8 ppm, respectively.

(14) Assume that the NO_x calibration curve for this NO-NO_x analyzer is a 45° straight line, i.e., the response is equal to the true concentration of NO_x input.

(15) Plot the true concentration of NO_x on *y*-axis vs NO_2 concentration on *x*-axis, and draw the converter efficiency curve.

(16) Since NO_2 input is converted to NO by the converter and detected as NO_x concentration, the slope of the curve times 100 is the average converter efficiency of the NO-NO_x analyzer.

(17) Determine the converter efficiency and discuss the result. Derive Equation 3.7.

EXPERIMENT 3—PREPARATION OF NO_2 STATIC TEST ATMOSPHERE

In this experiment, you will generate a static test atmosphere of NO_2 in a nonrigid chamber by adding a known amount of pure NO_2 from a cylinder to the nonrigid container containing a known volume of air.

Since high concentrations of pure NO_2 are subject to dimerizing according to the reversible reaction,

$$2NO_2 \rightarrow N_2O_4$$

The pure NO_2 in the cylinder will contain both NO_2 and N_2O_4. You will determine the appropriate volume of the NO_2/N_2O_4 mix to inject in order to obtain a specified concentration of NO_2 by examining the ratio of NO_2:N_2O_4 at laboratory temperature and pressure.

After the test atmosphere is generated, it will be analyzed by the NO-NO_x analyzer.

Procedure to follow is:

(1) Calculate the volume of the NO_2/N_2O_4 mix necessary to generate 8 ppm of NO_2 in the following manner:
   - Determine the percent by volume of N_2O_4 in a NO_2/N_2O_4 mix at room temperature and pressure from Figure 3.8.

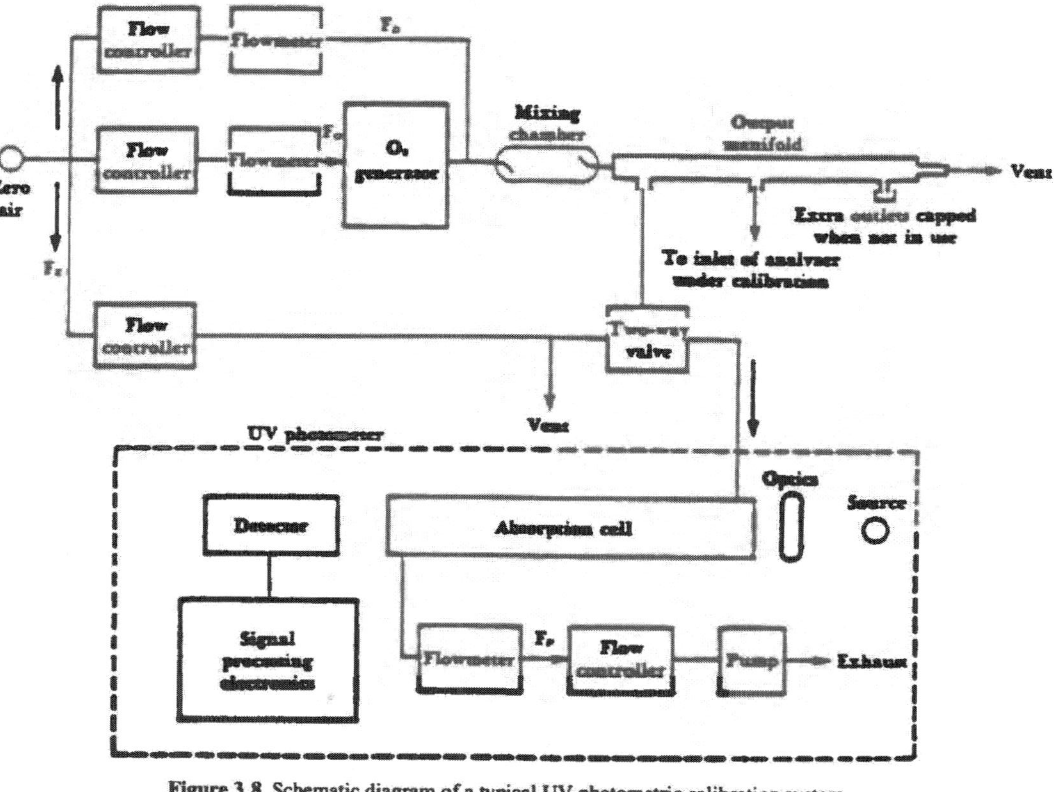

Figure 3.8 Schematic diagram of a typical UV photometric calibration system.

- Calculate the volume percent of NO_2 in the NO_2/N_2O_4 mix according to the equation

$$\text{vol \% } NO_2 = \left\{1 - \frac{\text{vol \% } N_2O_4}{100}\right\} 100 \tag{3.8}$$

- Calculate the injection volume of the mix according to the equation

$$V_I = \frac{cV_T}{2\left\{\frac{\text{vol \% } N_2O_4}{100}\right\} + \left\{\frac{\text{vol \% } NO_2}{100}\right\}} \tag{3.9}$$

where:

V_I = volume of injected NO_2 (μl)
V_T = total volume of the test atmosphere, 100 l
c = concentration of NO_2 in test atmosphere, ppm
vol % N_2O_4 = % by volume of N_2O_4 in the NO_2/N_2O_4 mix
vol % NO_2 = % by volume of NO_2 in the NO_2/N_2O_4 mix

(2) Evacuate the bag completely, utilizing the vacuum pump provided.
(3) Purge the bag twice, using the bag filling apparatus.
  - Adjust the needle valve (attached at the top of the rotameter) for a flow rate of approximately 10 liters per minute as measured on the rotameter.
  - Fill the bag at this rate for about 5 minutes.
  - Evacuate the bag completely, as in Step 2.
(4) Set the bag filling train for a flow rate of 10 liters per minute.
(5) Connect the bag and start the timer.
(6) After the bag has been filling for about 5 minutes, inject the computed volume of the NO_2/N_2O_4 mix
  - Insert the needle directly into the discharge tube of the NO_2 cylinder.
  - Fill and purge the syringe at least six times to ensure complete removal of any air from the syringe barrel and needle.
  - Fill the syringe to a volume greater than the required volume.
  - Avoid touching the syringe barrel or needle with your hand, as your body heat will change the density of gas in the syringe.
  - Immediately prior to injection, adjust the syringe to the required volume.
  - Insert the needle directly into the center of the septum on the bag filling rig. The needle should be perpendicular to the face of the septum at all times. Exercise care to avoid bending the needle.

- Withdraw the syringe immediately after injection to prevent additional pollutant from diffusing from the needle into the test atmosphere.

(7) Continuing filling the bag to a total volume of 100 liters (10 minutes at 10 liters per minute).
(8) Disconnect the bag from the filling rig and cap the bag.
(9) Turn off the air source.
(10) Connect the bag to the inlet of the NO-NO_x analyzer, and record the NO_x response.
(11) Discuss the result. Compare it with the result from Experiment 2.

OPERATION OF THE "THERMO ELECTRON" CHEMILUMINESCENT NO/NO_x ANALYZER

Zero Adjust and NO Calibration

(1) With all electrical power off, connect lines for the "oxygen" from a compressed air cylinder. Adjust the cylinder pressure regulator to deliver air at about 10 psi.
(2) Adjust the internal oxygen pressure to +2 psi ± 0.5 on the oxygen pressure gauge (located at the upper right side of the analyzer).
(3) Set the temperature controller on the NO_2-to-NO converter to 650°C and place the "INPUT MODE" switch to NO position.
(4) Turn on the vacuum pump and bypass pump.
(5) Turn on main AC power at control unit with "NO_x CONV" switch OFF.
(6) Observe the reaction chamber pressure gauge located at the very top of the analyzer unit. The chamber pressure gauge should indicate 5−12 mm Hg.
(7) Place the "RANGE PPM" selector switch on the control unit in the 2.5 ppm position.
(8) With the "SUPPRESSION" knob turned to the extreme counter-clockwise position, observe the dark current from the photomultiplier on the display meter. A reading of less than 10 ppm is typical, depending upon room temperature and photomultiplier characteristics.
(9) Place the "RANGE" selector switch in the "ZERO" and then the "FULL" positions, and observe correct indication on the control unit display meter.
(10) Return to the 2.5 or 10 ppm range and observe dark current reading.

If stable (a slight fluctuation is normal), turn background suppression potentiometer to set the meter on zero.

(11) Check the sample vacuum gauge on the upper left front of the analyzer unit for −5 inches Hg. If incorrect, adjust the regulator located behind the plug marked "SAMPLE."

(12) Turn on the "OZONE GENERATOR" switch on the analyzer unit.

(13) Set the range selector switch to obtain a near full reading with the NO standard gas (the NO concentration in the cylinder).

(14) Connect the sample line to the sample inlet.

(15) Adjust the pressure regulator at the NO cylinder and/or sample line to obtain 2 SCFH flow on the sample bypass flowmeter on the analyzer unit.

(16) Adjust the calibration dial on the control unit to obtain the correct reading for the NO calibration gas.

(17) Observe the instrument responses for several additional concentrations of NO input in order to draw the NO calibration curve.

NO_x Calibration

(18) Permit the zero air sample to flow through the system for 5 minutes.

(19) Switch the instrument to the NO_x mode by placing the "INPUT MODE" switch (on the converter) to NO_x position and turning on the "NO_x CONV" switch (on the control unit).

(20) Observe the converter cycle lamp on the temperature controller for cyclic operation. If the lamp cycles on-off, then the converter is at operating temperature. Usually 15–30 minutes is required for the converter to reach temperature and stabilize.

(21) Repeat Steps 11–17 as described for NO calibration.

OZONE IN THE ATMOSPHERE—EXPERIMENT

The measurement principle used here is chemiluminescence with ethylene. The calibration procedure is ultraviolet (UV) photometry.

LABORATORY OBJECTIVES

Upon completion of this laboratory, you should be able to:

(1) Properly use the Reference Measurement Principle for the determination of ozone in the ambient air and be able to calibrate an ozone instrument using the reference calibration procedure – UV Photometry.

(2) Describe the basic principles involved in ozone measurement by gas phase chemiluminescence with ethylene and by ultraviolet photometry.
(3) Recall the major components and their functions in the setup for the UV photometric calibration procedure.
(4) Describe the difference between a UV photometer used as an analyzer and a UV photometer used as an absolute photometer calibrator.
(5) Determine the following instrument performance parameters:
   - rise time
   - fall time
   - lag time
   - linearity

PRINCIPLE OF MEASUREMENT

The reference method for the measurement of ambient ozone is the C_2H_4-chemiluminescence method. (See *Federal Register*, Vol. 41, No. 826, Appendix F, 12-1-76: "Measurement Principle and Calibration Procedure for the Measurement of Nitrogen Dioxide in the Atmosphere.") Chemiluminescence is a term describing chemical reactions which emit light energy. A sample stream of ambient air and user-supplied ethylene are delivered simultaneously to a reaction chamber where the O_3 in the air reacts with the ethylene. The light energy emitted from this reaction is then measured by a photomultiplier tube which converts the light energy to an electrical signal.

The resulting current, which is related to the O_3 concentration, is amplified and either read directly or displayed on a recorder.

CALIBRATION OF OZONE MONITOR

Ozone monitors may be calibrated either by the boric acid potassium iodide (BAKI) method, or ultraviolet (UV) photometry. In both calibration methods, an ozone generator is used to produce constant concentrations of ozone. Mixtures containing ppm concentrations of O_3 in air are obtained by exposing air or oxygen to a mercury vapor lamp emitting short wavelength (185) radiation. Ozone formation results from the photodissociation of oxygen, and recombination with the molecular oxygen:

$$O_2 + hv \rightarrow 2O$$

$$O + O_2 \rightarrow O_3$$

The amount of O_3 produced is a function of the oxygen content of the air stream, the irradiating energy (i.e., the intensity of the lamp) and the flow rate of air. The O_3 concentration from the generator is then determined by using either the BAKI or UV methods. Several known O_3 concentrations generated from the ozone generator are then used to calibrate the ozone monitor.

UV photometry is the preferred method for the calibration of ozone monitors replacing the former BAKI calibration procedure. However, the BAKI method is a good example of the wet chemistry techniques that are used for the calibration of many air sampling instruments.

PROCEDURE

Using the procedures described in Title 40, Part 50 (Appendix D) of the U.S. Code of Federal Regulations, record data on Tables 3.8 and 3.9.

General

(1) Check and adjust all flow controllers and instrument gauges for proper flow rates and operation parameters.

(2) Calculate proper span number for Dasibi 1003-AH, and dial this into the instrument' s electronics.

(3) Perform a five-point calibration of the chemiluminescence ozone analyzer using the Dasibi calibrator according to the procedure in the TAD *Federal Register*, Vol 44, No. 28, Appendix D, 2-8-79: "Measurement Principle and Calibration Procedure for the Measurement of Ozone in the Atmosphere."

(4) Determine rise time, fall time and lag time for the chemiluminescence ozone analyzer.

(5) Determine linearity for the Dasibi UV photometer.

Calibration Preparations

(1) Check to see if both the chemiluminescence ozone analyzer and the Dasibi UV photometer are on and operating normally.

(2) Set the sample +ethylene flow on the ozone analyzer to the flow rate setting specified by the lab instructor. Record this setting on the calibration data sheet.

(3) Set the ethylene flow rate to that specified by the lab instructor. Record this setting.

(4) Set the Dasibi UV photometer at a flow rate of 2 liters/min (2000 cc/min) on the flow rate gauge at the front of the instrument. Record this setting.

TABLE 3.8. Laboratory—Ozone Calibration Data Sheet.

Name________________ Group #__________ Date__________

Analyzer serial number__________
UV photometer serial number__________
Ozone generator number__________

Calibration preparations

Analyzer sample and ethylene flow rate__________
Analyzer ethylene flow rate__________
Dasibi sample flow rate__________
Calibration system flow rate__________
Dasibi sample frequency__________
Dasibi control frequency__________
Dasibi 1008-PC temperature __________
Dasibi 1008-PC pressure__________
Dasibi zero value__________ (Remember to subtract this from all Dasibi concentration readings)

Dasibi preparations

Detector temperature (T)__________K
Barometric pressure (P_b)__________Torrs
Calculate the new span number (N span) using the following formula:

$$N_{span} = \frac{10^6}{(308)71} \times \frac{760}{P_b} \times \frac{T}{273}$$

New span number__________

Note: Extra steps are required if you are using a Dasibi 1003

TABLE 3.9. Laboratory—Ozone Linearity Check Data Sheet.

Name________________ Analyzer serial no. ____________
Group no. ________________ Model no. ____________
Date________________
Ozone generator flow F_o (L/min)________________

Dilution flow F_d L/min	Dilution ratio (R)	Average of three photometer readings (ppm)	Linearity error %E
0	1		

$$\%E = \frac{A_1 - \frac{A_2}{R}}{A_1} \times 100 \qquad R = \frac{F_o}{F_o + F_d}$$

Where: E = linearity error in percent
A_1 = assay of original concentration
A_2 = assay of diluted concentration
R = dilution ratio = flow of original concentration divided by total flow

(5) Check the flow rate gauge at the initial needle valve (just upstream of the ozone generator) to be sure that you have enough flow through the calibration system. This should be approximately 1–2 liters/min higher than required by the Dasibi and ozone monitor (sum of their two flow rates). The lab instructor will give you a suggested flow setting. Record this setting.

(6) Record the sample frequency and control frequency readouts for the Dasibi photometer. If you are using the Dasibi 1008-PC photometer, also record the temperature and pressure readouts.

(7) Record the Dasibi zero value. This is done by switching the function switch to the N span position and subtracting N span from the number on the readout panel (e.g., 46.850 readout − 46.800 N span = .050 zero offset). *This number must be subtracted from all Dasibi concentration readouts to get the actual ozone concentrations.*

(8) Set the ozone analyzer so that it measures in the 0-.5 ppm range (.5 ppm will be full scale).

(9) Zero the ozone analyzer using the zero pot dial. Lock the pot after zeroing.

Calibration Procedure

(1) Turn the ozone generator on and pull the sleeve out at least 20 mm. Allow the generator to warm up and equilibrate for 10 minutes.

(2) Slide the ozone generator's sleeve in or out until the Dasibi's concentration readout indicates an ozone level of approximately .45 ppm. *Note*: Allow the Dasibi to cycle at least six time before determining the ozone concentration.

(3) Allow six more cycles to pass and record concentration readings from the Dasibi and average these readouts. Record this average concentration reading.

(4) Adjust the ozone analyzer's span control until the analyzer's readout matches the Dasibi's average ozone concentration. Record the analyzer concentration readout.

(5) Adjust the ozone generator sleeve to obtain another ozone concentration (e.g., .40 ppm). Wait six cycles, take six cycles' readings, record the average Dasibi reading, record the ozone analyzer reading.

(6) Repeat the above steps until you have at least five ozone concentrations covering the entire 0–.5 ppm range (e.g., .45, .35, .25, .15, .05).

(7) Construct a calibration curve for the ozone analyzer using the data you've recorded above. Apply linear regression to achieve the line of best fit. Hand in worksheet on calibration curve drawn.

Determination of Instrument Performance Parameters

Determine rise time, fall time and lag time for the chemiluminescence ozone monitor. Perform test procedures Steps 18–29 of Appendix A, Section 5, and complete the necessary calculations. Record all final results on the Instrument Performance Data Sheet.

Ultraviolet (UV) Photometer Linearity Check Procedure

(1) Assemble apparatus as in Figure 3.8.

(2) Generate an ozone concentration of approximately 0.45 ppm, and determine flow rate (F_o) necessary to achieve this concentration (ob-

tained from flow meter calibration curve). Record on linearity check data sheet.

(3) Reduce original concentration to approximately 0.40 ppm by dilution with zero air. Record dilution flow rate (F_d) and measured ozone concentration.

(4) Generate ozone concentrations by dilution of approximately 0.30, 0.20, 0.10 ppm and record respective (F_d) and measured ozone concentrations.

(5) Calculate the dilution ratio (R) for each concentration generated using the following equation.

$$R = \frac{F_o}{F_o + F_d}$$

Record on linearity check data sheet.

(6) Calculate percent linearity error for each concentration generated using the following equation.

$$\%E = \frac{A_1 - \frac{A_2}{R}}{A_1} \times 100$$

where:

E = linearity error in percent
A_1 = assay of original concentration
A_2 = assay of diluted concentration
R = dilution ratio = flow of original concentration divided by total flow

Record $\%E$ on linearity check data sheet.

If the linearity error exceeds 5%, the accuracy of the flow dilution should be checked and verified carefully before the photometer is assumed to be inaccurate. The test should be carried out several times at various dilution ratios, with an averaging technique used to determine the final result. If any modifications to the UV system are necessary in performing the linearity test, care should be exercised to avoid introducing leaks or other adverse effects.

If the linearity error is excessive and cannot be attributed to flow measurement inaccuracy, then the photometer system should be checked for:

- dirty or contaminated cell, lines or manifold
- inadequate "conditioning" of system
- leaking two-way valve or other leak in system

- contaminant in zero air
- nonlinear detectors in the photometer
- faulty electronics in the photometer

Also, UV system nonlinearity might be indicated when a nonlinear calibration curve is obtained for an analyzer that is expected to be linear.

METHANE IN AIR—EXPERIMENT

INTRODUCTION

(1) Methane (partition mode)—A gas chromatograph with flame ionization detection separates methane from other hydrocarbons. The detector response is calibrated against methane of known concentration.
(2) Total hydrocarbon (THC mode)—Total hydrocarbons are determined using the gas chromatograph without a packed column—all hydrocarbons are detected unseparated. The detector response is a function of the total number of carbon atoms present. This response is calibrated against methane of known concentration.

RANGE AND LIMIT OF DETECTION

(1) The range of the method for the analysis of methane is from approximately 0.01 ppm to approximately 20,000 ppm.
(2) The minimum detectable concentration in the analysis of methane is approximately 0.01 ppm.

INTERFERENCES

(1) Any compound which has the same retention time as methane at the operating conditions described in this method is a potential interferent in the methane determination. However, no interferences are known to exist.

PRECISION AND ACCURACY

(1) The precision of the method when a fixed loop gas sampling valve is used for sample injection is ±3%.
(2) The accuracy of the method has not been determined. It is planned to reference the cylinders to Standard Reference Materials (SRM) for methane when such standards are again available from the National Bureau of Standards (NBS). However, it would not be an SRM as NBS would not have a history of the cylinder.

ADVANTAGES AND DISADVANTAGES OF THE METHOD

(1) The advantage of the method is that all hydrocarbon span gas cylinders in the statewide air monitoring network will be traceable to one set of secondary standard span gases, thereby improving accuracy and reproducibility.
(2) The disadvantage of the method is that at the present time, there is not a standard traceable to the National Bureau of Standards.

APPARATUS

(1) GC equipped with a FID, a 1.0 ml fixed loop gas sampling valve, and a switching valve capable of switching from a chromatograph column in the partition mode of operation to no column in the THC mode of operation
(2) Gas chromatograph column – Any column capable of separating methane from other hydrocarbons in the partition mode. A stainless steel column with 17% *b,b'*-oxydipropionitrile stationary phase on 60/80 mesh activated alumina solid support has been used successfully
(3) Strip chart recorder with a one millivolt range and one second response is acceptable. A mechanical or electronic integrator or a computer-based data system is also recommended.
(4) Tedlar bags, 12" × 15", for the preparation of standards
(5) 250 μl gas-tight syringe for preparation of standards
(6) 5 ml gas-tight syringe fitted with a shutoff valve
(7) Calibrated flowmeter is required to accurately measure the flow rate of the diluent air in the preparation of standards
(8) Dilution apparatus for the preparation of standards
(9) Timer

REAGENTS

(1) *Carrier Gas* – Compressed synethetic (zero) air, nitrogen or helium prepared from water pumped sources of guaranteed high purity
(2) *Hydrogen* – High quality electrolytic generators which produce ultrapure hydrogen under pressure are recommended. They provide some degree of safety when compared to the explosive hazard of bottled hydrogen. All connections must be made with thoroughly cleaned stainless steel tubing and properly tested for leaks. If hydrogen

in compressed gas cylinder is used, the hydrogen is prepared from water pumped sources.

(3) *Oxygen* –Compressed oxygen, 99.6 mole percent

(4) *Methane* –99.97 Mole percent

(5) *Synthetic (Zero) Air* –Hydrocarbon free

PROCEDURE

(1) Analysis of Cylinder Samples

- G.C. conditions–Establish operating conditions for the G.C. according to manufacturer's recommendations, so that methane has a retention time of about 0.5 minutes. Set the switching valve of the G.C. for the partition mode of operation (G.C. column on stream). Set the carrier flow, hydrogen and oxygen flow rates at the manufacturer's recommended flow rates. These are nominally 25, 100 and 300 ml/min, respectively. Descriptions of the operation of FIDs have been published and should be consulted. If at all possible, maintain room environment within ±0.5°C.
- Partition analysis–Flush the 5 ml syringe several times with the gas from the cylinder to be analyzed. Shut the valve on the syringe, and remove the syringe from the gas cylinder. Attached the syringe to the inlet port of the gas sampling valve of the G.C. Open the shut-off valve of the syringe, and flush the contents through the gas sampling valve and sample loop of the G.C. Introduce the sample into the G.C. by turning the sample injection valve. Five (5) replicate analyses of each sample should be made.
- Measurement of area–The area of the sample peak is measured by an electronic integrator, computer data system, or some other suitable form of area measurement such as peak height *x* width at half height. Calculate concentrations as described in the "Calculations" section.
- Total hydrocarbon analysis–Set the switching valve to the THC mode of operation (no column on stream) and the flow rates as described in the first item under step 1 so that methane has a retention time of about 0.1 minutes. Repeat the procedure in the second and third items under Step 1.

CALIBRATION AND STANDARDS

(1) Bag quality control–Flush five (5) 12" × 15" Tedlar bags five (5) times each with synthetic air. Fill the bags with synthetic air and

analyze for the total hydrocarbon content. Check for leaks by immersing in water. Allow the bags to stand overnight and reanalyze the next day. Any change in the THC may be indicative of a problem from the bag. Discard bad bags.

(2) Preparation of standards

- Evacuate the five (5) bags using house vacuum or vacuum pump, and prepare the appropriate concentration (9, 15 or 45 ppm) of methane in synthetic air diluent.
- Connect the dilution apparatus to the tank of synthetic air and set flow to 1 liter/min with the calibrated flowmeter. Connect the opened valve of a Tedlar bag to the dilution apparatus with a short length (1") of plastic tubing. Turn the three-way stopcock to begin filling the bag and start the timer. Withdraw predetermined volume of methane from a flowing stream from the standard cylinder of methane. When about two liters of synthetic air are in the bag, inject this amount of methane through the septum of the glass tee of the dilution apparatus, and allow it to be flushed into the bag. Continue filling the bag to five (5) liters. When five liters are in the bag, rotate the three-way stopcock to shut off flow to the bag; and close the valve on the bag. Remove the bag from the dilution apparatus, and knead the bag briefly (~1 min). Repeat this procedure until five (5) bags have been prepared at the predetermined concentration.

(3) Calibration of G.C.

- Partition analysis—Chromatograph the contents of each bag in the partition mode. Perform five (5) replicate analyses on each bag.
- Measurement of area—Measure the area of the standard peaks as was done in the first item under Step 1 in "Procedure" section.
- Flush and fill five bags with the synthetic air used as diluent. Chromatograph each bag five (5) times. Measure the peak areas as in the first item under Step 1 in "Procedure" section.
- Total hydrocarbon analysis—Set the switching valve to the THC mode of operation. Repeat the procedure in the previous 3 items.

  Note: Bag standards must be analyzed at the same time that the sample analysis is done. This will minimize the effect of day-to-day variations and variations during the same day of the FID response. Bag standards should be made fresh each time they are required for standardization of AIHL Secondary Span Gases.

CALCULATIONS

(1) Determine the mean area response of the 25 measurements of standard in the partition mode. Subtract the mean area response of the synthetic air diluent in the partition mode. Calculate the response factor:

$$R_f = \frac{C}{A_{std} - A_{air}}$$

where:

R_f = response factor in ppm/mm^2 or ppm/count
C = concentration of standard in ppm
A_{std} = area in mm^2 or counts of standard plus air
A_{air} = area in mm^2 or counts of diluent air

(2) Using the response factor determined in Step 1, calculate the concentration for each analysis of the sample span gas cylinder and the mean concentration and standard deviation of the measurements for the cylinder.

(3) Repeat the calculations as in Step 1 and Step 2 for the measurements made in the THC mode.

OTHER AIHL SECONDARY STANDARDS

(1) Repeat steps in the previous 3 sections for the other two cylinders that will remain at AIHL, preparing bag standards at the appropriate cylinder concentrations.

ARB Working Standards

(1) Partition analysis
- Calibration of G.C.—Perform five (5) replicate calibrations of the G.C. with an AIHL Secondary Standard. Calculate the mean response factor as in Step 1 under "Calculations."
- Analysis of ARB working standard—Perform five (5) replicate analyses of the appropriate Working Span Gas Standard. Calculate the mean concentration and the standard deviation of the measurements as in Step 2 under "Calculations."
- Repeat the procedure in the previous 2 items for the other two gas concentrations.

(2) THC analysis
- Calibration of G.C.—Set the switching valve to the THC mode

of operation as in the last item in "Procedure" section. Repeat the procedure in the previous 3 items.

UNIQUE AEROSOL SAMPLING

BIOAEROSOL SAMPLERS

Introduction

Bioaerosols are a loosely defined group of airborne particles of variable biological origin, i.e., viruses, bacteria, fungal spores, pollen and various antigens. The need for their accurate sampling is of increasing importance, especially in indoor air studies, outdoor allergen monitoring and in occupational exposure situations. There are currently several samplers available which utilize various principles of operation and therefore have different design and collection characteristics.

The physics of particle removal from an airborne state, as well as principles of good sample collection, are common to all airborne particles. The basic concepts concerning the process of air sampling to identify and quantify airborne contaminants are well known in industrial hygiene and can be applied to the sampling of bioaerosols. Two key elements of any air sampling process are sample collection and sample timing. Efficient removal of various size microbes from the air and their collection into or onto a medium for identification is based on the physical characteristics of the airborne particles and on the physical parameters of the sampler. Hence, for example, a sampler designed to mimic the collection efficiency of human lungs would not collect large (>10 μm) pollen grains. Sampling for an optimal time that truly represents the environment is a statistical challenge that must balance the variation with time of the ambient concentrations with the characteristics of the sampling device. For example, a sampler designed to sample a large volume of air over a long period of time in a clean environment would quickly become overloaded in an agricultural situation. An optimal sampling time is especially problematic for bioaerosols which are known to have vast temporal changes in concentrations.

Collection Process

Bioaerosol sampling involves separating the particle trajectory from the air streamline trajectory. To achieve this, different physical forces are used as illustrated in Figure 3.9. In Figure 3.9(a), the inertia of the particle forces its impaction onto a solid or semi-solid surface, usually either a culture medium, or an adhesive that can be examined microscopically.

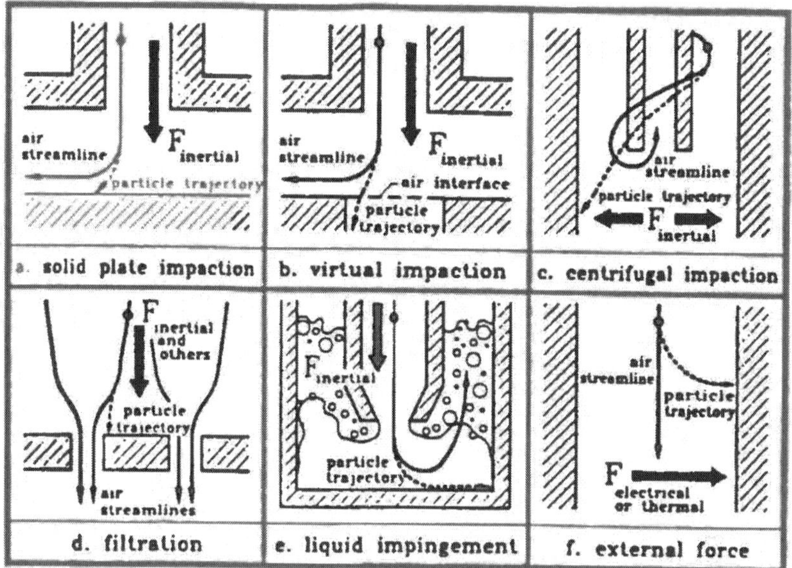

Figure 3.9 Mechanisms of particle removal from air.

Figure 3.9(b) illustrates a virtual impactor, which is also based on the inertial behavior of particles. Size separation of particles occurs when the small lower-inertia particles flow with the air streams and the large high-inertia particles cross the "virtual" impaction surface and are collected or sensed below this interface. Such a sampler, also referred to as a dichotomous sampler because of the split into two flows, has been used to collect outdoor air allergens.

The principle of particle separation by centrifugal force is shown in Figure 3.9(c). It also uses the inertial behavior of the particle, but in a radial geometry.

When filtration is the collection mechanism, as shown in Figure 3.9(d), inertial forces are also responsible separating the particles from the air stream. However, in concert with inertial impaction, other mechanisms, such as interception, diffusion and electrostatic attraction contribute to the deposition of particles onto the filter material. Liquid impingement, Figure 3.9(e), is a method that mainly uses inertial forces to collect particles, but also uses diffusion within the bubbles to enhance particle collection. Illustrated in Figure 3.9(f), this principle is widely used for the size measurement of aerosols, but it has not yet gained commercial status for bioaerosol sampling. Some samplers of this type currently exist in experimental stages and will probably have field-use applications in the future.

Collection Time

Aerosol concentrations vary greatly with time; this applies especially to bioaerosols and their sources. An example could be the release of millions or billions of fungal spores which may occur abruptly due to wind or other source disturbances causing ambient concentration variations of several orders of magnitude.

General Comments

Collection efficiency is a measure of how well the sampler deposits the particles without being affected by their physical properties. Biological efficiency refers to the ability of the sampler to maintain the microbial viability and prevent cell damage during sampling. Other problems not associated with sampling efficiency can create sampling bias. Statistical bias in quantitative analyses which results from inappropriate sample surface densities is related to inappropriate sampling procedures, such as unsuitable sampler selection or improper sampling periods. Similarly, problems associated with the cultivation of the sampled microbes are not properties of the sampler, but rather a problem of the analysis phase. In order to quantify the total effect of all these efficiency factors, they must be analyzed separately. Only after all these effects have been well characterized can bioaerosol samplers be properly utilized and their results used to explain the nature and behavior of bioaerosols.

AEROSOL COLLECTING OF MULTI-LAYER SNOW-COVERED SAMPLES

This instrument is shown in Figure 3.10.

This is a unique instrument intended for collecting snow layers of 2 centimeters thickness with a volume (V_i) of about 1000 cm^3. The number of snow layers collected depends on total snow-cover height and number of incorporated collecting stages. This device can be used for determination of total dust, alkaline and heavy metals concentration. However, for this the snow container must be made of nonmetal material. The device is also useful for snow density determinations. In practice, snow samples are collected from three or more smooth places to eliminate random errors.

Snow density and concentration of contaminating substances in snow are expressed as follows:

$$D_i = \frac{M_i}{V_i} \tag{3.10a}$$

Figure 3.10 Instrument for collecting multi-layer snow-covered aerosol samples.

$$C_{i,k} = \frac{m_{i,k}}{M_i} \tag{3.10b}$$

$$C_{i,k} = \frac{m_{i,k}}{V_i} \tag{3.10c}$$

where:

D_i = snow density in layer i
M_i = the mass of molten snow in layer i
$C_{i,k}$ = the concentration of contaminating substance k in snow layer i
$m_{i,k}$ = mass of contaminating substance k in snow layer i

This instrument is especially useful in field measurements. It is a simple, small sampling device. It is possible to use it to study the profiles of snow contaminations as a function of time.

CHAPTER 4

Source Sampling

INTRODUCTION

SOURCE-SAMPLING PURPOSES

SOURCE sampling or emissions testing, as applied to air pollution studies, is the procedure whereby a representative sample is removed from some larger, contaminant-bearing gas stream confined in a duct or stack. This sample is then subjected to further analysis, and the contaminant concentrations are related to the parent gas stream to determine total quantities. Because the sample extracted from the main gas stream usually represents a very small fraction of the total volume, extreme care should be exercised in obtaining a representative sample. Additionally, because of the many and variable factors encountered in sampling gas streams, complex methods must frequently be used to obtain representative samples.

Source sampling frequently is employed to answer a variety of questions of which the main one is: What are the quantities and concentrations of emissions? Subsequent questions that can be answered from the basic determination include:

(1) Is the process in compliance with present or expected emission regulations?

(2) What is the efficiency of existing pollution control equipment?

(3) What effect do various process variables have on emissions?

(4) Is a valuable product or byproduct being emitted?

(5) What are the potential (uncontrolled) emissions of various processes?

LEGAL USE OF SOURCE-SAMPLING INFORMATION

Every test should be conducted as if it were ultimately to be used as evidence in court. The collection and analysis of source samples should become a routine matter to the agency personnel involved. It must be remembered, however, that this routine procedure is too esoteric for the layman and therefore subject to greater scrutiny whenever the agency has to rely on its results. It is imperative that source sampling and analysis be done under standard procedures and that each step be well documented. In short, the report may ultimately be subjected to the requirements of the Rules of Evidence.

This chapter will discuss the standardization of source-sampling procedures relative to taking the sample, chain of custody, laboratory analysis, report custody, and disposition of the original worksheets.

Taking the Sample

In attacking the validity of source-sampling results, the adverse party will concentrate on four main items relative to taking the sample: (1) sampling procedure, (2) recorded data and calculations, (3) test equipment, and (4) qualifications of the testing personnel.

Agency personnel must be aware of the possibility of adverse inferences that may arise from the use of unorthodox or new procedures. Thus deviations from the standard procedure must be kept to a minimum and applied only where absolutely necessary to obtain an accurate sample. Changes in methodology must be based on sound engineering judgment and must be carefully documented. Standard procedures that should receive particular attention are:

(1) Location of sampling station
(2) Number and size of sampling zones in the duct
(3) Use of recommended sampling equipment
(4) Careful determination of gas velocities
(5) Maintenance of isokinetic sampling conditions
(6) Proper handling of the collected sample and recording of container and filter numbers

Close scrutiny will also be focused upon the recorded field data, because these data form part of the physical evidence. Standardized forms should be utilized to ensure that there is no lack of necessary information. Forms designed for this purpose should include field, laboratory, and calculation forms. Only the field form is utilized when taking the sample. This form is designed to identify clearly the process tested, the date and time, location of test station, sampling personnel, and the person who recorded the data. During the actual test period, the meter readings, temperature readings, and other pertinent data should be recorded in the provided spaces immediately upon observation. These data determine the accuracy of the test and should not be erased or altered. Any errors should be crossed out with a single line, and the correct value should be recorded above the crossed-out number.

Test Equipment

Faulty test equipment can also invalidate a test. In general, there are two types of field test equipment, gas-sampling and process-measuring equipment.

The process-measuring equipment consists of any of the metering devices by which test data are obtained. These devices include scales for weighing fuel or raw materials and orifices and gauges for measuring product flow. Because proper maintenance and calibration procedures are often lacking, it cannot be assumed that these devices are accurate. In any case, check and record the date on which the devices were serviced.

Ideally, the use of process-measuring equipment should be kept to a minimum. Process-weight regulations, however, may frequently require the use of such equipment, especially scales. Such scales can only be properly serviced and calibrated by specially trained personnel. The scale manufacturer usually provides this service. A stamp affixed to the scale by the service crew as a standard procedure will note the date of calibration or inspection. If the scale has not been recently calibrated, an engineering judgment must be made concerning its accuracy. A material balance will sometimes provide a check on scale readings.

PROCEDURE FOR LEAK CHECKING SOURCE SAMPLING TRAINS

PURPOSE

To ensure that the sample collected is representative of the source, the sampling train must be leak free. All sampling trains used by or on behalf of California Air Resources Board (ARB) for the collection of any samples must be checked for leaks before and after each test. The following are some suggested procedures for performing these leak checks.

ASSEMBLING TRAIN COMPONENTS

At the testing site the sampling train components are assembled carefully to ensure that all connections are air tight and leak-free. All ground-glass ball and socket connections must be clamped together with positive-lock pinch clamps. A light coating of silicone grease is applied to the outer portion of the male ground-glass joints. If Tygon rubber or Teflon tubing is used for connecting impingers or other components, the internal diameter of the flexible tubing should be one-eighth of an inch less than the outer diameter of the mating fixture.

Sampling equipment mounted in a mobile monitoring van is particularly vulnerable to vibrations. In this case, all sample line connections should be rigid screw type, otherwise screw clamps which compress an airtight seal at all connections must be used.

Spring clamps and clips may not be used to secure connections on a mobile installation unless the items being connected are securely mounted or clamped in place so as to exclude the imposition of tension on the joint.

LEAK CHECKS

Prior to use in the field individual train components (e.g., meters, pumps) should be checked for internal leaks by applying air pressure. All leak checks of the assembled train or monitoring system must be performed at the test site. Both positive (pressure) and negative (vacuum) leak checks should be performed, if applicable. When the pump is located at the rear end of the train, the entire train is operated under a vacuum. Occasionally, the pump is located immediately ahead of the dry gas meter, in which case the sample lines, ahead of the pump are under vacuum and the section between the pump and meter is under positive pressure. The pressure portion of the train may easily be checked by the application of air pressure. For most applications, the vacuum leak check is the primary check which must be performed on the assembled train.

Leak checks are performed just prior to and immediately following the completion of each test. A vacuum leak check on an EPA type integrated sampling train should be performed as follows:

(1) Plug intake nozzle with an airtight stopper.

(2) Turn on the pump.

(3) Turn the coarse adjustment valve on the console (meter box) to the "off" position and open the fine-adjust valve until the vacuum gauge reads 15 inches of mercury. After the dial has stabilized, the flow rate should stop. If a flow rate in excess of 0.01 cubic feet during 30 seconds is observed, the leak or leaks must be found and corrected.

(4) If the leak rate is found to be satisfactory, the plug is first slowly removed from the nozzle after which the pump is turned off. This procedure prevents water from the impingers from being forced backward into the filter.

METHOD 5 FOR PARTICULATES—EXPERIMENT

INTRODUCTION

U.S. Environmental Protection Agency Reference Method 5 ("Determination of Particulate Emissions from Stationary Sources"), collectively with Methods 1, 2, 3 and 4, comprises the most widely used system for evaluating particulate emissions from stationary sources. This method involves extraction of a gas sample isokinetically from the stack. Particulate matter is collected on an out-of-stack glass fiber filter maintained at 120° ± 14°C (248 ± 25°F) or at a temperature specified by an applicable subpart of the standards or approved by the administrator. The mass of particulate matter, which includes any material that condenses at or above the specified filter temperature, is measured gravimetrically.

As opposed to some methods, the most significant error associated with this test method occurs during the sample collection and recovery phase instead of in the analysis phase. Therefore, this method requires careful attention to the entire procedures to get an accurate result.

Objectives of this work are to understand the concept of isokinetic sampling and to determine the particulate concentration and emission rate of the stationary source using Reference Method 5.

ISOKINETIC SAMPLING (SEE ALSO TABLE 4.1)

Isokinetic sampling conditions exist when the velocity of the gases entering the probe nozzle tip (v_n) is exactly equal to the velocity of the approaching stack gases (v_s), that is $v_n = v_s$. The percent isokinetic is defined as:

$$\% \text{ isokinetic} = \frac{v_n}{v_s} \times 100$$

and is equal to 100% only when $v_n = v_s$. When $v_n \neq v_s$ (anisokinetic conditions), sample concentrations can be biased due to the inertial effects of particles.

If the gas-flow streamlines are disturbed as in anisokinetic conditions:

(1) Large particles tend to move in the same initial direction.
(2) Small particles tend to follow the streamlines.
(3) Intermediate particles are somewhat deflected.

EPA SAMPLING TRAIN AND METHOD 5 HIGHLIGHTS

The EPA sampling train is illustrated in Figure 4.1. A sharp-edged stainless

TABLE 4.1. Isokinetic Rate Equation.

Simplified Isokinetic Rate Equation

$$\Delta H = K\Delta p$$

Nozzle Tip Volumetric Flow Rate

$$Q_n = A_n v_n = A_n v_s = \frac{\pi D_n^{\,2}}{4} v_s$$

Orifice Meter Equation

$$Q_m = K_m \sqrt{\frac{T_m \Delta H}{P_m M_m}}$$

T and P Correction for Dry Gas Stream

$$Q_n = \frac{P_m}{P_s}\frac{T_s}{T_m} Q_m$$

Moisture Correction

$$n_s(1 - B_{ws}) = n_m(1 - B_{wm})$$

Flow Rate Corrected for T, P and Moisture

$$Q_n = \frac{1 - B_{wm}}{1 - B_{ws}}\frac{T_s}{T_m}\frac{P_m}{P_s} Q_m$$

Relation of Flow Rate at Nozzle to Meter Flow Rate

$$Q_n = \frac{(1 - B_{wm})\, T_s P_m}{(1 - B_{ws})\, T_m P_s} K_m \sqrt{\frac{T_m \Delta H}{P_m M_m}}$$

$$\frac{\pi D_n^{\,2}}{4} v_s = \frac{(1 - B_{wm})\, T_s}{(1 - B_{ws})\, T_m}\frac{P_m}{P_s} K_m \sqrt{\frac{T_m \Delta H}{P_m M_m}}$$

Pitot Tube Equation

$$v_s = K_p C_p \sqrt{\frac{T_s \Delta p}{P_s M_s}}$$

TABLE 4.1. (continued).

$$\frac{\pi D_n^2}{4} K_pC_p \sqrt{\frac{T_s\Delta p}{P_s M_s}} = \frac{(1 - B_{wm})}{(1 - B_{ws})} \frac{T_s}{T_m} \frac{P_m}{P_s} K_m \sqrt{\frac{T_m \Delta H}{P_m M_m}}$$

Solving for ΔH

$$\Delta H = \left\{ D_n^4 \left(\frac{\pi K_pC_p}{4 K_m}\right)^2 \frac{(1 - B_{ws})^2}{(1 - B_{wm})^2} \frac{M_m}{M_s} \frac{T_mP_s}{T_sP_m} \right\} \Delta p$$

Moisture Relationships

$$M_m = M_d(1 - B_{wm}) + 18B_{wm}$$

$$M_s = M_d(1 - B_{ws}) + 18B_{ws}$$

Isokinetic Rate Equation

$$\Delta H = \left\{ D_n^4 \left(\frac{\pi K_pC_p}{4 K_m}\right)^2 \frac{(1 - B_{ws})^2}{(1 - B_{wm})^2} \frac{[M_d(1 - B_{wm}) + 18B_{wm}]T_mP_s}{[M_d(1 - B_{ws}) + 18B_{ws}]T_sP_m} \right\} \Delta p$$

$\Delta H_@$ is defined as the orifice pressure differential that gives 0.75 CFM of air at 68°F and 29.92" Hg.

$$\Delta H = \frac{Q_m^2}{K_m^2} \frac{P_mM_m}{T_m}$$

$$\Delta H_@ = \frac{(.75 \text{ CFM})^2 (29.92'' \text{ Hg}) (29.0)}{(460 + 68) K_m^2}$$

$$\Delta H_@ = \frac{.9244}{K_m^2}$$

Simplifying:

assume $B_{wm} = 0$

let $\Delta H_@ = \dfrac{.9244}{(K_m)^2}$

and $K_p = 85.49$

Isokinetic Rate Equation—Working Form

$$\Delta H = \left\{ 846.72\, D_n^4\, \Delta H_@ C_p^2 (1 - B_{ws})^2 \frac{M_d}{M_s} \frac{T_m}{T_s} \frac{P_s}{P_m} \right\} \Delta p$$

TABLE 4.1. (continued).

Nozzle Diameter Selection (for units of: D_n in inches, Q_m in scfm, B_{ws} in fraction, T_s in °R, P_s in inches of Hg, and Δp in inches of water)
$$D_n = \sqrt{\left(\frac{.0358 Q_m P_m}{T_m C_p (1 - B_{ws})}\right)\left(\frac{T_s M_s}{P_s (\Delta p)}\right)}$$

NOMENCLATURE

A_n = sampling nozzle cross-sectional area, A_s = stack cross-sectional area, a = mean particle projected area, B_{wm} = fraction moisture present in gas at meter, B_{ws} = fraction moisture present in stack gas, C_p = pitot tube calibration coefficient, $C_{p(std)}$ = standard pitot-static tube calibration coefficient, c_s = particulate concentration in stack gas mass/volume, c_{s12} = particulate concentration corrected to 12% CO_2, c_{s50} = particulate concentration corrected to 50% excess air, D_E = equivalent diameter, D_H = hydraulic diameter, D_n = source sampling nozzle diameter, E = emission mass/heat Btu input, e = base of natural logarithms (ln 10 = 2.302585), %EA = percent excess air, F_c = F-factor using c_s and CO_2 on wet or dry basis, F_d = F-factor using c_s and O_2 on a dry basis, F_w = F-factor using c_{ws} and O_2 on a wet basis, F_o = miscellaneous F-factor for checking Orsat data, $\Delta H_@$ = pressure drop across orifice meter for 0.75 CFM flow rate at standard conditions, ΔH = pressure drop across orifice meter, j = equal area centroid, K_p = pitot tube equation dimensional constant:

$$\text{Metric units} = 34.97\ \text{m/sec.}\left[\frac{\text{g/g-mole (mmHg)}}{(°\text{K})(\text{mm H}_2\text{O})}\right]^{1/2}$$

$$\text{English units} = 85.49\ \text{ft./sec.}\left[\frac{\text{lb/lb-mole (in. Hg)}}{(°\text{R})(\text{in. H}_2\text{O})}\right]^{1/2}$$

L = length of duct cross section at sampling site, ℓ = path length, L_1 = plume exit diameter, L_2 = stack diameter, m = mass, M_d = dry stack gas molecular weight, M_s = wet stack gas molecular weight, n = number of particles, N_{Re} = Reynolds number, O_1 = plume opacity at exit, O_2 = in stack plume opacity, P_{atm} = atmospheric pressure, P_b = barometric pressure ($P_b = P_{atm}$), P_m = absolute pressure at the meter, pmr = pollutant mass rate, P_s = absolute pressure in the stack, P_{std} = standard absolute pressure—metric units = 760 mm Hg, English units = 29.92 in. Hg, Δp = gas velocity pressure, $\Delta p_{(std)}$ = standard velocity pressure read by the standard pitot tube, Δp_{test} = gas velocity pressure read by the type "S" pitot tube, q = particle extinction coefficient, Q_s = stack gas volumetric flow rate corrected to standard conditions, R = gas law constant, 21.83(in. Hg)(ft.^{3})/(lb − mole)(°R), t = temperature (°Fahrenheit or °Celsius), T_m = absolute temperature at the meter—metric units = °C + 273 = °K, English units = °F + 460 = °R, T_s = absolute temperature of stack gas, T_{std} = standard absolute temperature, metric units = °20°C + 273 = 293°K, English Units = 68°F + 460 = 528°R, V_m = volume metered at actual conditions, $V_{m\,std}$ = volume metered corrected to standard conditions, v.p. = water vapor pressure, v_s = stack gas velocity, Volume H_2O = metric units = 0.00134 m^3/ml × ml H_2O, English units = 0.0472 ft.^{3}/ml × ml H_2O, W = width of the duct cross section at the sampling site, θ = time in minutes, Subscripts: atm = atmospheric, ave = average, b = barometric, d = dry gas basis, f = final, g = gauge, i = initial, m = at meter, n = at nozzle, p = of pitot tube, s = at stack, SCF = standard cubic feet, std = standard conditions, w = wet basis.

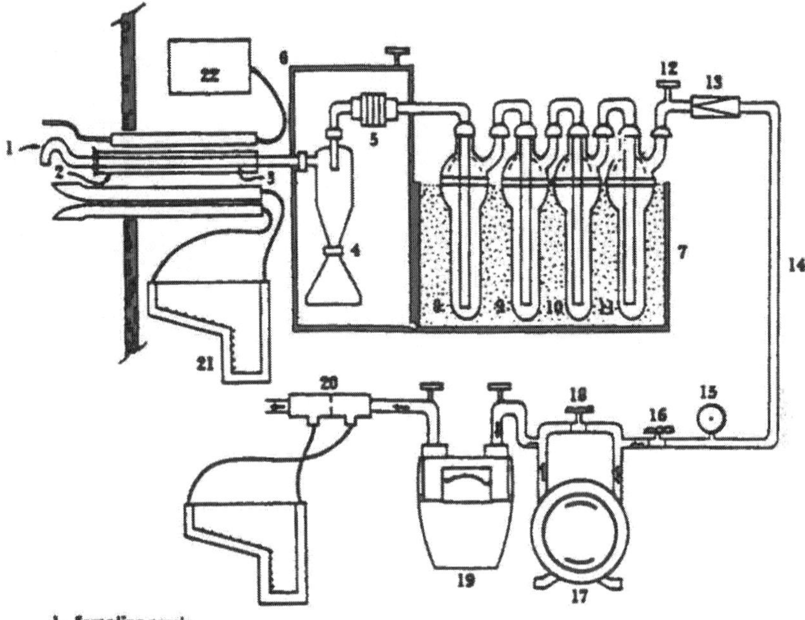

1. Sampling nozzle
2. Sampling probe sheath
3. Heated sample probe liner
4. Cyclone assembly (proposed regulations do not require this cyclone)
5. Out of stack filter assembly
6. Heated filter compartment maintained 120 °C ± 14 °C (248 °F ± 25 °F) (or temperature specified in 40CFR subpart)
7. Impinger case
8. First impinger filled with H_2O (100 ml)
9. Greenburg-Smith (or modified Greenburg-Smith) impinger filled with H_2O (100 ml)
10. Third impinger – dry
11. Fourth impinger – filled with H_2O absorption media (200-300 gm)
12. Impinger exit gas thermometer
13. Check valve to prevent back pressure
14. Umbilical cord – vacuum line
15. Pressure gage
16. Coarse adjustment valve
17. Leak free pump
18. By-pass valve
19. Dry gas meter with inlet and outlet dry gas meter thermometer
20. Orifice meter with manometer
21. Type S pitot tube with manometer
22. Stack temperature sensor

Figure 4.1 EPA Method 5 particulate sampling train.

steel nozzle is used with a heated borosilicate or quartz glass-lined or stainless steel sample probe. A type S pitot tube is attached directly to the sample probe to allow simultaneous sample collection and velocity measurement.

The particulate collection medium is specified as a glass fiber filter, without organic binder, exhibiting at least 99.95% efficiency (<0.05% penetration) on 0.3-micron dioctyl phthalate smoke particles. The filter is placed against a glass frit filter support with a silicone rubber gasket in a heated filter holder. This is followed by an impinger train or a measuring condenser, then an air-tight pump, dry gas meter and orifice meter to measure the flow rate.

The number and positions of sampling points are determined from reference Method 1. A total sampling time (no. of traverse points × sampling time per point) is selected to be greater than or equal to the minimum total sampling time specified in the test procedures for the specific industry so that:

(1) The sampling time per point is ≥2 min.
(2) The sample volume corrected to standard conditions exceeds the required minimum total gas sample volume.

Before sampling, the equipment is leak-checked by plugging the nozzle and pulling a 15 in. Hg vacuum. A leak rate of not more than 0.02 cfm is considered acceptable.

At the completion of the test, the pump is turned off (the pump is turned off any time the probe is removed from the port, but not during traversing in the same port) and the probe and nozzle removed from the stack. Post-test leak check is again required at the conclusion of each sampling run with the same procedure as of the pretest leak check. The equipment is then removed to a clean location and disassembled. Particulate matter collected in the nozzle, probe and all parts prior to the filter is washed out with acetone and kept in a sealed container. The water volume or weight collected in the condenser is measured for the moisture determination. The filter and acetone washings are analyzed gravimetrically for particulate weight.

PROCEDURE

Presampling Preparation

(1) Select an appropriate nozzle size according to your preliminary data.
(2) Be sure to get $\Delta H_{@}$, dry gas meter calibration factor Y, and an orifice meter calibration curve (plot between ΔH across orifice and sample flow rate) from the instructor.
(3) Determine required total sampling time according to the following procedure:

- Calculate expected ΔH (pressure drop across the orifice which has to be adjusted according to Δp in the stack in order to get an isokinetic sampling rate) from the isokinetic rate equation.

$$\Delta H = \left\{ 846.72\, D_n^4\, \Delta H_{@} C_p^2 (1 - B_{ws})^2 \frac{M_d}{M_s} \frac{T_m}{T_s} \frac{P_s}{P_m} \right\} p$$

where:

ΔH = pressure drop across the orifice meter, in. H_2O
D_n = selected nozzle diameter, in.
$\Delta H_{@}$ = pressure drop across orifice meter for 0.75 CFM flow rate at standard conditions, in. H_2O

- Determine the sample flow rate corresponding to the calculated ΔH from the orifice meter calibration curve.
- The total sampling time must be selected from the greater value of the following: (a) minimum total sampling time specified for the specific industry (in the NSPS regulations), and (b) required minimum total gas sample volume corrected to stack conditions divided by the expected sample flow rate from Step 2.
  Note: For this exercise, use: (a) minimum total sampling time = 40 min. and (b) minimum gas sample volume = 30 scf.
- The sampling time per traverse point is selected from the greater value of the following: (a) 2 min., and (b) total sampling time from (a) divided by number of required traverse points.
  Note: It is recommended that the number of minutes sampled at each point be a whole minute or a whole and one-half minute, in order to avoid timekeeping errors. The sampling time at each point must be the same.

(4) Number on the reverse and desiccate the filters at 20 ± 5.6°C (68 ± 10°F) and ambient pressure for at least 24 hours. Record the weight to the nearest 0.1 mg.

(5) Number and weigh six 250 ml or 300 ml beakers to the nearest 0.1 mg, and keep them in a clean, dry place.

(6) Prepare the collection train according to 40 CFR 60 (Appendix A) (see Appendix of this manual). It is recommended that both water volume and the weight of water or silica gel plus impinger be measured and entered in the attached data forms.

(7) Conduct the pretest leak check according to 40 CFR 60 except that the leak check is done before the filter and probe heating systems are turned on. (Be sure to understand it all thoroughly before you start.)

(8) Leak check the pitot tube according to the procedure described in Chapter 1.

(9) Read through the particulate field data form provided. See how many readings you have to take at each traverse point, then organize among your team to make all of these readings. One person has to be responsible for Δp readings and the ΔH adjustment at the meter box.

Sampling Operation

(1) Record the initial dry gas meter readings, barometric pressure and other data and enter in the field data form provided.

(2) Mark the probe according to the required traverse point positions.

(3) Position the tip of the probe at the first sampling point with the nozzle tip pointing directly into the gas stream.

(4) When in position, block off the open area around the probe in the porthole to prevent flow disturbances and unrepresentative dilution of the gas stream.

(5) Turn on the pump and immediately adjust the sample flow rate, i.e., adjust ΔH, to attain isokinetic conditions according to the velocity pressure (Δp) at that point. Record both Δp and adjusted ΔH.

Note: It is recommended that a calculator program be used for rapid determination of the orifice pressure drop (ΔH) corresponding to the isokinetic sampling rate. Reset your program if the absolute stack temperature (T_s) changes more than 10%. More than one adjustment of ΔH may be necessary if the velocity pressure (Δp) is fluctuating substantially at that point (>20% variation). Be sure to record every pair of Δp and ΔH that occur during your sampling.

(6) Take the other readings required in the field data form.

(7) Record the dry gas meter readings at the end of each time interval (sampling time at each point).

(8) Repeat Steps 3 through 7 for each sampling point.

Note: Move the tip of the probe to the next sampling point in the same port without turning the pump off. You may turn the pump off during the test only when you finish with one port and want to remove the probe from the stack and put it into another port.

(9) When you have traversed through all the points, turn off the pump, remove the probe from the stack and record the final readings.

(10) Turn off the filter and probe heating system.

(11) Conduct the post-test leak check according to 40 CFR 60.

(12) Leak-check the pitot lines.

(13) Disconnect the probe and sampling train.

Note: Periodically during the test, observe the connecting glassware from the probe through the filter to the first impinger, for water condensation. If any condensation is evident, adjust the probe and/or filter heater setting upward until the condensation is eliminated; add ice around the impingers to maintain the silica gel exit temperature at 20°C (68°F). The manometer level and zero should also be checked periodically since vibration and temperature fluctuations can cause the manometer zero to shift.

Sample Recovery and Analysis

(1) Filter – Carefully remove the filter from the filter holder, and place it in its designated petri dish. Any filter fibers or particulates which adhere to the filter gasket should be removed with a nylon bristle brush or a sharp blade and placed in the same container. Note the color of the particulate collected. Desiccate the filter for 24 hours and weigh to a constant weight to the nearest 0.1 mg.

(2) Probe and connecting glassware (acetone rinse) – Initially put 100 ml of acetone n the acetone rinse sample bottle.

Clean the outside of the probe, the pitot tube and the nozzle to prevent particulates from being brushed into the sample bottle.

Carefully remove the probe nozzle, and rinse the inside surface (using a nylon bristle brush and several acetone rinses) into the sample bottle until no particles are visible in the rinse.

Clean inside of the swagelok fitting by the same procedure described in (2)

Use at least two people to rinse the probe, using the following procedure:

- Rinse the probe liner by tilting and rotating the probe while squirting acetone into the upper (or nozzle) end to ensure complete wetting of the inside surface.
- Allow the acetone to drain into the sample bottle using a funnel to prevent spillage.
- Hold the probe in an inclined position and squirt acetone into the upper end while pushing the probe brush through the liner with a twisting motion, and catch the drainage in the sampler bottle.
- Repeat the brushing procedure three or more times until no particles are visible in the wash or until a visual inspection of the liner reveals no remaining particulate matter inside.
- Rinse the liner once more.

- Rinse the brush to collect any particulates which may be retained within the bristles.
- Wipe all the connecting joints clean of silicone grease, and clean the inside of the front half of the filter holder by rubbing the surface with a nylon bristle brush and rinsing it with acetone.

  Repeat the procedure at least three times or until no particles are evident in the rinse.
- Make a final rinse of the filter holder and brush.
- Clean any connecting glassware which precedes the filter holder using the same procedure as above.

Use a dry, clean glass funnel to transfer the acetone rinse into the dry, clean 250 ml graduated cylinder.

Record the volume of the sample to the nearest 1.0 ml and transfer it into a dry, clean, tared and numbered 250 or 300 ml beaker. Use the data form provided.

Repeat the previous two steps until you finish with the acetone rinse sample. Usually, the acetone rinse sample will be about 500–700ml.

Rinse the container with two or three 25 ml portions of acetone, cap the container and shake, then transfer into the graduated cylinder to rinse it, and then put the rinse through the funnel into a beaker; thus, the container, the graduated cylinder and the funnel have been rinsed.

Note: Be sure to make a record of every ml of acetone you use.

Put 100 ml of acetone in a beaker to be evaluated as a blank (in order to determine the residue in the acetone reagent itself).

Let the sampler and blank dry at no more than 100°F in a dust-free environment or under a watchglass.

Transfer the totally evaporated samples and blank into a tightly sealed desiccator that contains fresh drierite.

Desiccate for 24 hrs and then weigh.

(3) Impingers:
- Make a notation on the sample recovery form of any color or film in the impinger water and if the color of the indicating silica gel indicates it has been completely spent.
- Wipe all the connecting joints clean of silicone grease and weigh all four impingers to the nearest 0.5 g.
- Transfer the collected water in each impinger to a graduated cylinder to measure water volume.

Calculation

Use the provided calculation sheet to determine the following:

(1) Moisture content

(2) Gas velocity and volumetric flow rate (using Orsat data)
(3) Particulate concentration (lb/dscf), emission rate (lb/hr) and percent isokinetic

Report

The report must at least include the following:

(1) Sampling time calculation
(2) Raw data (field and analysis)
(3) All calculations
(4) Summary of results
(5) Discussion and conclusions

METHODS 1–4 SOURCE SAMPLING

INTRODUCTION

Stationary source sampling is a procedure for evaluating the characteristics of an industrial waste gas stream in a duct or stack. Contaminants in the air from these sources may be solid, liquid or gas; organic or inorganic. The procedures outlined in the Code of Federal Regulations, Methods 1–5 for isokinetic stationary source sampling are a versatile system for evaluating these pollutant concentrations.

The sampling system measures a number of variables at the source while extracting from the gas stream a sample of known volume. The information on source parameters in conjunction with quantitative and qualitative laboratory analysis of the extracted sample makes possible calculation of the total amount of pollutant material in the gas stream.

Isokinetic source sampling provides a great deal of important data on the operational variables and emissions of an industrial stationary source. This information is used as the basis for decisions on a variety of issues. The data taken during a source test experiment must, therefore, be a precise representation of the real situation. This task requires a thorough knowledge of the recommended sampling procedures in conjunction with an understanding of process operations, thus making source testing an endeavor that should be performed only by trained personnel.

Objectives are:

(1) To determine the number of traverse points for a particulate traverse (Reference Method 1)
(2) To determine the molecular weight of the gas sample using an Orsat Analyzer (Reference Method 3)

(3) To determine the moisture content of the gas stream using the wet bulb-dry bulb technique (alternative of the approximation method mentioned in Reference Method 4)
(4) To determine gas stream velocity and volumetric flow rate (Reference Method 2)
(5) To determine an appropriate size of sampling nozzle for Isokinetic Sampling

Experiment 1—Traverse Points Determination

In order to get a representative measurement of pollutant concentration and/or total volumetric flow rate from a stationary source, a measurement site where the gas stream is flowing in a known direction is selected, and the cross section of the duct or stack is divided into a number of equal *areas*. A traverse point is then located within each of these equal areas.

Procedure

DETERMINING THE NUMBER OF TRAVERSE POINTS

(1) Measure the inside dimensions of the duct at the sampling site. Record this data in.
(2) If the duct has a rectangular cross section, calculate the duct equivalent diameter according to the equation:

$$D_e = \frac{2LW}{L + W}$$

where:
D_e = Equivalent diameter of the duct
L = Length of the duct cross section
W = Width of the duct cross section

(3) Measure the distance from the sampling site to the nearest downstream flow disturbance (distance B in Figure 4.2).
(4) Divide these distances by the equivalent diameter of the duct.
(5) Determine the corresponding number of traverse points for *each* distance from Figure 4.2. This number must be a multiple of two.
(6) Select the higher of these two numbers. This is the minimum number of traverse points that must be used for preliminary velocity measurement and particulate sampling.

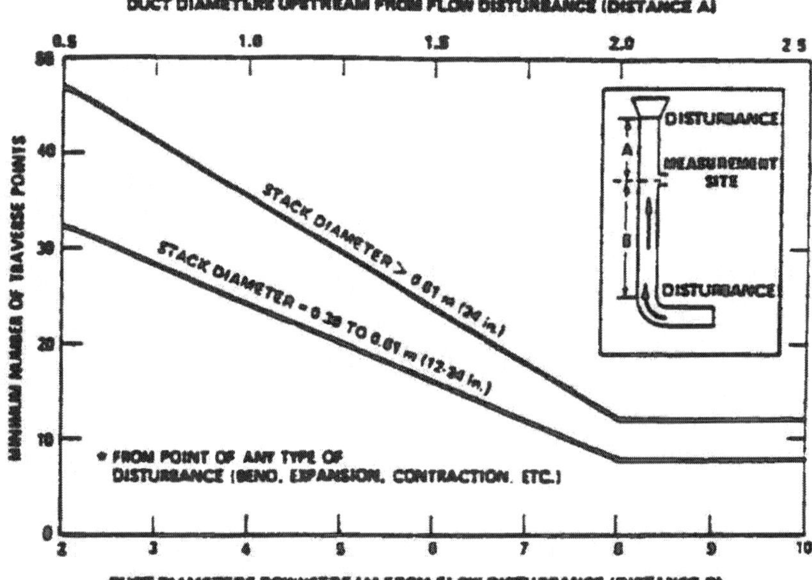

Figure 4.2 Minimum number of traverse points.

CROSS-SECTIONED LAYOUT AND LOCATION OF TRAVERSE POINTS

(1) For circular ducts, locate the traverse points on two perpendicular diameters according to Table 4.2.

(2) For rectangular ducts, divide the duct cross section into as many equal rectangular areas as there are traverse points. Maintain the length-to-width ratio of the areas between 1.0 and 2.0. Then, locate a traverse point at the center of each individual area (see Figure 4.3).

(3) Draw a diagram showing the sampling site cross section and traverse points location.

Experiment 2—Dry Gas Molecular Weight Determination

Reference Method 3 is to be used for determining the excess air and dry molecular weight of gas streams from fossil-fuel combustion processes. It is also applicable to other processes if compounds other than CO_2, O_2, CO and N_2 are not present in sufficient concentrations to affect the results.

In this experiment, a gas sample will be extracted from the duct using a single-point grab sampling method (one sampling point at the centroid of

TABLE 4.2. **Percent of Stack Diameter from Inside Wall to Traverse Point.**

Traverse Point Number on a Diameter[a]	Number of Traverse Points on a Diameter[b]											
	2	4	6	8	10	12	14	16	18	20	22	24
1	14.6	6.7	4.4	3.2	2.6	2.1	1.8	1.6	1.4	1.3	1.1	1.1
2	85.4	25.0	14.6	10.5	8.2	6.7	5.7	4.9	4.4	3.9	3.5	3.2
3		75.0	29.6	19.4	14.6	11.8	9.9	8.5	7.5	6.7	6.0	5.5
4		93.3	70.4	32.3	22.6	17.7	14.6	12.5	10.9	9.7	8.7	7.9
5			85.4	67.7	34.2	25.0	20.1	16.9	14.6	12.9	11.6	10.5
6			95.6	80.6	65.8	35.6	26.9	22.0	18.8	16.5	14.6	13.2
7				89.5	77.4	64.4	36.6	28.3	23.6	20.4	18.0	16.1
8				96.8	85.4	75.0	63.4	37.5	29.6	25.0	21.8	19.4
9					91.8	82.3	73.1	62.5	38.2	30.6	26.2	23.0
10					97.4	88.2	79.9	71.7	61.8	38.8	31.5	27.2
11						93.3	85.4	78.0	70.4	61.2	39.3	32.3
12						97.9	90.1	83.1	76.4	69.4	60.7	39.8
13							94.3	87.5	81.2	75.0	68.5	60.2
14							98.2	91.5	85.4	79.6	73.8	67.7
15								95.1	89.1	83.5	78.2	72.8
16								98.4	92.5	87.1	82.0	77.0
17									95.6	90.3	85.4	80.6
18									98.6	93.3	88.4	83.9
19										96.1	91.3	86.8
20										98.7	94.0	89.5
21											96.5	92.1
22											98.9	94.5
23												96.8
24												98.9

[a]Points numberec from outside wall toward opposite wall

[b]The total number of points along two diameters would be twice the number along a single diameter.

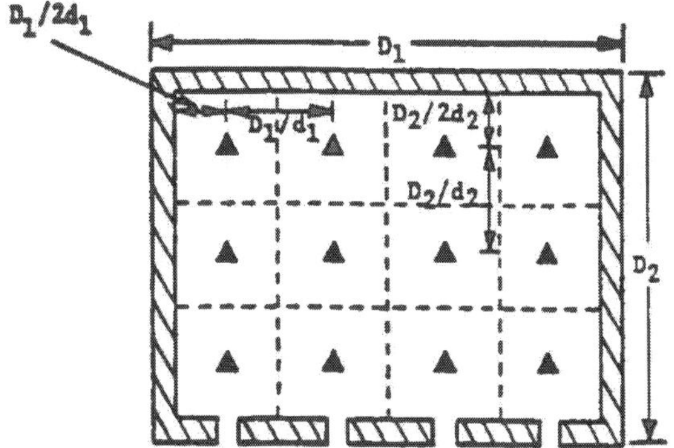

where:

▲ = sampling point

d_1 = number of areas across flue width

d_2 = number of areas across flue perpendicular to width

Figure 4.3 Example showing rectangular stack cross section divided into twelve equal areas, with a traverse point at the centroid of each area.

the duct cross section) and analyzed for percent CO_2, percent O_2 and percent CO using an Orsat analyzer (see Figure 4.4).

Procedure

LEAK-CHECK PROCEDURE

(1) Allow the apparatus to reach ambient temperature with the manifold valve open and the three pipette valves closed.

(2) Bring the liquid in each absorption pipette up to the reference mark by opening the pipette valves one at a time and by slowly lowering the leveling bottle. Pinch off the rubber tube to the leveling bottle with the heel of the hand to quickly stop the liquid flow. Close the pipette valves.

(3) Displace the indicating fluid until a reading is obtained in the narrow part of the burette, and quickly close the manifold inlet valve.

(4) Place the leveling bottle on top of the Orsat case, and read the meniscus in the burette.

(5) Wait at least 4 minutes, then read the meniscus again.

(6) A change of ≥ 0.2 ml in the reading indicated a leak in the system which must be repaired. A drop in reagent level to below the capillary tube over a 4-minute period indicates a leak in that pipette.

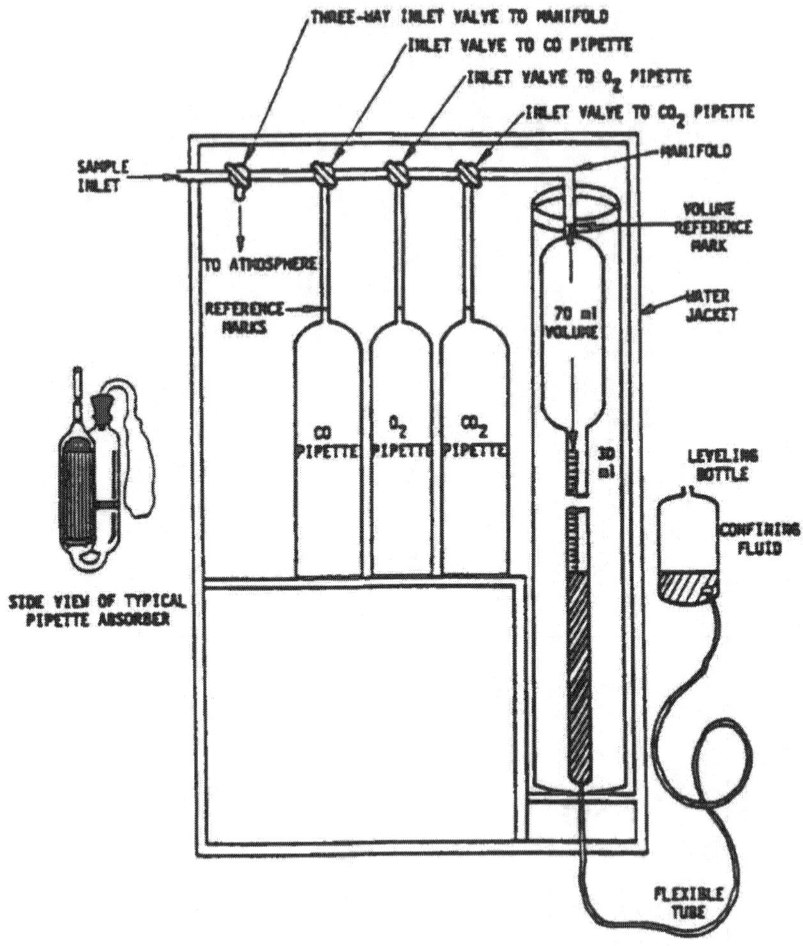

Figure 4.4 Orsat apparatus.

(7) If leaks are detected, correct them and repeat Steps 2–5.

ORSAT SAMPLING

(1) Close all inlet valves to the absorption pipettes.
(2) Open the three-way inlet valve to the manifold.
(3) Lower the leveling bottle until all the indicating fluid flows out of the glass burette.
(4) Use the squeeze bulb to draw a gas sample from the duct into the glass burette. About 10 pumps of the bulb are required to replace the air previously confined in the burette with new gas sample.

(5) Switch the three-way inlet valve to open to the atmosphere (not to the sampling port).

(6) Carefully and slowly raise the bottle, letting the indicating liquid flow into the burette until the volume of gas sample is reduced to 10 ml at atmospheric pressure by pushing the excess gas sample out through the three-way valve to the atmosphere. Note that the gas sample volume is at atmospheric pressure when the liquid in the leveling bottle is level with the indicating liquid in the burette.

(7) Close the three-way valve when the indicating liquid reaches zero. Now, you have 100 ml of gas sample in the glass burette, and it is ready to be analyzed.

ORSAT ANALYSIS

(1) With the liquid in each absorption pipette at the reference mask and the three-way valve closed, open the inlet valve to the CO_2 pipette while moving the leveling bottle upward, letting the gas sample flow into the CO_2 absorption pipette.

(2) Keep on raising the leveling bottle carefully and slowly until the indicating liquid reaches the volume reference mask on the glass burette, then stop.

(3) Lower the bottle to pull back the gas sample from the absorbing solution. Slowly and carefully move it downward until the CO_2 absorbing solution reaches the reference mark at the CO_2 pipette. *Do not let the solution go into the manifold.* Read the new volume of gas in the burette with the leveling bottle held level with the liquid in the burette.

(4) Repeat raising and lowering the bottle until you obtain two equal readings (about 5 passes).

(5) Take a final reading of the remaining volume of gas sample after CO_2 absorption by closing the inlet valve to the CO_2 pipette (after the absorbing solution has been brought back to the reference mark) and adjusting the leveling bottle to the same level of liquid in the graduated burette. (This will bring the gas pressure in the burette to atmospheric pressure which is the same as the initial condition when you adjust the gas sample volume to 100 ml.)

(6) Read the meniscus for the volume of CO_2 in the gas sample.

(7) Repeat Steps 1–6 for the other two pipettes. Perform about 100 passes for O_2 and 5 passes for CO.

(8) Record the data on the data sheet (Table 4.3).

(9) The dry molecular weight of the gas sample ($M_{s,dry}$) is calculated from the equation:

TABLE 4.3. Dry Molecular Weight Determination.

PLANT ______________________________ COMMENTS:

DATE ____________ TEST NO ____________

SAMPLING TIME (24-hr CLOCK) ____________

SAMPLING LOCATION ____________

SAMPLE TYPE (BAG, INTEGRATED, CONTINUOUS) ____________

ANALYTICAL METHOD ____________

AMBIENT TEMPERATURE ____________

OPERATOR ____________

ORSAT LEAK CHECKED ____________

RUN / GAS	1		2		3		AVERAGE NET VOLUME	MULTIPLIER	MOLECULAR WEIGHT OF STACK GAS (DRY BASIS) M_d, lb/lb-mole
	ACTUAL READING	NET	ACTUAL READING	NET	ACTUAL READING	NET			
CO_2								44/100	
O_2 (NET IS ACTUAL O_2 READING MINUS ACTUAL CO_2 READING)								32/100	
CO (NET IS ACTUAL CO READING MINUS ACTUAL O_2 READING)								28/100	
N_2 (NET IS 100 MINUS ACTUAL CO READING)								28/100	
								TOTAL	

$$M_d = .44\ (\%CO_2) + .32(\%O_2) + .28(\%CO + N_2)$$

$$M_{s,dry} = (\%CO_2)(0.44) + (\%O_2)(0.32) + (\%N_2)(0.28)$$

(10) The wet molecular weight is then:

$$M_{s,wet} = (M_{s,dry})\left(\frac{100 - \%H_2O}{100}\right) + (\%H_2O)(0.18)$$

Note: There is no error counting CO as N_2 as the molecular weights are the same.

Experiment 3—Moisture Determination

The stack gas moisture determination may be obtained by sampling a known volume of air and condensing the water vapors in an ice cooled condenser (as described in Reference Method 4) or by measurement of the wet and dry bulb temperatures of the stack gas. In practice, the former method is used in conjunction with the actual sampling, while the latter method may be used before testing to provide data for calculations to estimate the isokinetic sampling rate.

In this experiment, you will use the wet bulb-dry bulb technique to determine the moisture content in the gas stream. The wet and dry bulb method requires withdrawing the gas stream through a sample system provided with a psychrometer. By determining the wet and dry bulb temperatures, the moisture can be calculated from the equation:

$$B_{ws} = \frac{V.P.}{P_{abs}}$$

where:

B_{ws} = water vapor content in the gas stream, proportion by volume, dimensionless

$V.P.$ = vapor pressure of H_2O, in Hg
= $S.V.P. - (3.67 \times 10^{-4})(P_s)(t_d - t_w)[1 + (t_w - 32)/1571]$

$V.P.$ = saturated water vapor pressure at wet bulb temperature (inches of Hg)

P_s = absolute pressure of stack gas, in Hg

t_d = dry bulb temperature, °F

t_w = wet bulb temperature, °F

It is noted that the dry bulb rapidly reaches equilibrium, while the wet bulb rises to equilibrium, levels off, and then rises again once the wick becomes dry. The inflection point at which the temperature reaches equilibrium is considered the wet bulb temperature.

Procedure

Measure the wet and dry bulb temperatures and elapsed times in order to get the temperatures at equilibrium condition. Then, calculate the moisture content according to the above equation. (The absolute pressure of the gas will be obtained from Experiment 4.)

Experiment 4—Gas Velocity and Volumetric Flow Rate Determination

In this experiment, you will perform the Reference Method 2 to determine gas velocity and volumetric flow rate in the duct. A calibrated Type "S" ("stauscheibe" or "reverse" type) pitot tube will be used to measure velocity pressure of the gas stream. This pressure measurement, along with the gas molecular weight, static pressure and temperature is used to determine the velocity of the gas in the duct. The velocity, then, becomes the basic parameter necessary for volumetric flow rate calculation.

Procedure

PITOT TUBE AND DIFFERENTIAL PRESSURE GAUGE CHECK

(1) Connect the pitot tube to the differential pressure gauge with the pitot tube lines.
(2) Check for obstructions by flowing lightly on one pitot tube leg and then the other, watching the response of the gauge.
(3) Check for leaks by blowing into the upstream leg (the one facing the gas flow direction), sealing the opening, and noting any drop in the pressure gauge reading. Check the downstream leg (the one that measures static pressure) by drawing a slight vacuum, sealing and noting the gauge. If there are no leaks, the gauge readings will remain constant. No change in the differential pressure gauge reading should occur.

VELOCITY MEASUREMENT

(1) Leak-check the pitot tube and differential pressure gauge. Measure the barometric pressure.
(2) Mark the pitot tube according to the location of the traverse points determined in Experiment 1.
(3) Insert the pitot assembly into the duct to each traverse point.
(4) Seal the port.
(5) Measure the velocity pressure and temperature at the designated point. Record the data on the data form.

(6) Measure the static pressure in the duct at the centroid of the duct cross section by removing the velocity pressure line from the opening facing upstream from the differential pressure gauge.

CALCULATION

(1) The average gas velocity is calculated from the equation:

$$V_s = K_p C_p (\Delta p)_{avg}{}^{1/2} \left[\frac{(T_s)_{avg}}{P_s M_s}\right]^{1/2}$$

where:

V_s = average gas velocity in the duct, ft/sec
K_p = unit conversion factor (usually called pitot tube constant)
$= 85.49 \dfrac{\text{ft}}{\text{sec}} \dfrac{(\text{lb/lb} - \text{mole})(\text{in. Hg})}{(°\text{R})(\text{in. H}_2\text{O})}^{1/2}$

C_p = pitot tube coefficient, dimensionless (will be provided by the instructor)
$(\Delta p)_{avg}$ = average square root of each individual velocity head (Δp), (in. H_2O)$^{1/2}$
$(T_s)_{avg}$ = average absolute gas stream temperature, °R
P_s = absolute gas static pressure, in. Hg
= barometric pressure ± duct static pressure
M_s = gas molecular weight, wet basis, lb/lb-mole
$= M_d(1 - B_{ws}) + 18.0B_{ws}$
M_d = gas molecular weight, dry basis, lb/lb-mole
B_{ws} = water vapor content in the gas stream, dimensionless fraction

(2) The average dry standard volumetric flow rate of the gas stream is calculated from the equation:

$$Q_{std} = 3600(1 - B_{ws})V_s A \left[\frac{T_{std}}{(T_s)_{avg}}\right]\left[\frac{P_s}{P_{std}}\right]$$

where:

Q_{std} = average dry standard volumetric flow rate, dscf/h
B_{ws} = water vapor content, dimensionless fraction
V_s = average gas velocity, ft/sec
A = cross-sectional area of the duct, ft^2

T_{std} = standard absolute temperature = 528°R
$(T_s)_{avg}$ = average absolute gas stream temperature, °R
P_s = absolute gas static pressure, in. Hg
P_{std} = standard absolute pressure = 29.92 in. Hg

Experiment 5—Isokinetic Nozzle Size Determination

In order to sample at an isokinetic flow rate, the proper nozzle diameter can be calculated from the equation:

$$D_n = \sqrt{\left[\frac{0.0358 Q_m P_m}{T_m C_p (1 - B_{ws})}\right] \left[\frac{T_s M_s}{P_s (\Delta p)_{avg}}\right]^{1/2}}$$

where:

D_n = required nozzle diameter, inches
Q_m = required sampling flow rate = 0.75 CFM
P_m = expected absolute gas sample pressure at the dry gas meter, in. Hg (Use the barometric pressure for this value.)
T_m = expected absolute gas sample temperature at the dry gas meter, °R (Use the ambient temperature at the sampling location for this value.)
C_p = pitot tube coefficient, dimensionless
B_{ws} = water vapor content, dimensionless fraction
T_s = average absolute gas stream temperature, °R
M_s = gas molecular weight, wet basis, lb/lb-mole
P_s = absolute gas static pressure, in. Hg
$(\Delta p)_{avg}$ = average gas velocity pressure, in. H_2O

CHAPTER 5

Sampling and Analysis of Toxic Compounds in Air

SAMPLING METHODS

SAMPLING techniques for gas phase organic components are summarized in Table 5.1. Some of these procedures are discussed elsewhere in this manual. Others are discussed here. For quality assurance procedures, see 40 CFR, Ch 1, Part 58, Appendix A, 7-1-90 Edition: "Quality Assurance Requirements for State and Local Air Monitoring Stations (SLAMS)."

BACKGROUND—IMPORTANT CHEMICAL AND PHYSICAL PROPERTIES

In order to select optimal sampling techniques for a given compound or group of compounds, one must consider the important properties of the compound(s) of interest having an effect on the sampling process. Physical properties to be considered include boiling point, vapor pressure, water solubility and solubility in various organic solvents. If a compound has sufficiently low volatility to be associated with particulate matter, one must also consider the size distribution of the particles and heat of adsorption of the compounds of interest.

Chemical properties of concern include thermal stability, reactivity with water or other common constituents in ambient air, and photochemical reactivity. The dissociation constant (pK_a or pK_b) for ionizable compounds should also be considered. The exact manner in which the various chemical and physical properties impact the sampling scheme is discussed below, with the discussion of individual sampling methods.

There are several useful handbooks which can be consulted to obtain information concerning the chemical and physical properties of organic compounds. These include the *Handbook of Chemistry and Physics*, the *Merck Index*, the *Handbook of Environmental Data on Organic Chemicals*, "Dangerous Properties of Industrial Materials," by N. I. Sax, and the

TABLE 5.1. **Summary of Sampling Techniques.**

Solid adsorbents
—Organic polymers (Tenax, XAD-2)
—Inorganic (silica gel, Florisil)
—Carbon (activated carbon, carbon molecular sieves)
Cryogenic trapping
Impingers
Whole air collection (canisters, glass bulbs)
Derivatization techniques

"Chemical Emergency Preparedness Programs," USEPA92230-1A. Some of these also contain useful information on environmental effects and regulatory limits for various industrial chemicals.

METHODS FOR GAS PHASE COMPONENTS

Compounds which are predominantly in the gas phase at ambient temperature and pressure are generally sampled by passing the air sample through a filtration device to remove particulate material prior to capture of the gaseous components. The sampling techniques commonly used for gas phase components are summarized in Table 5.1. In selected cases, direct analysis of the filtered gas stream is possible, circumventing the need for the capture process (e.g., direct GC with photoionization or electron capture detection). This situation is rare in ambient air monitoring for toxic organics because the low concentrations generally make preconcentration of the sample a necessity.

Solid Adsorbents

Solid adsorbents are the media most commonly employed for sampling gas phase organics. The primary advantage of this sampling approach is the large volume of air which can be sampled relative to other techniques such as impingers or cryogenic sampling. These media can be generally divided into three categories as follows:

(1) Organic polymeric adsorbents
(2) Inorganic adsorbents
(3) Carbon adsorbents

Organic polymeric adsorbents include materials such as Tenax® GC and XAD-2. These materials have the important feature that water is not collected in the sampling process and hence large volumes of air can be collected. Another advantage of the organic polymeric adsorbents is the absence of "active sites" which can lead to irreversible adsorption of certain polar compounds. A major disadvantage of these materials is their inability to capture highly volatile materials (e.g., vinyl chloride) as well as certain polar materials (e.g., methanol, acetone).

Inorganic adsorbents include silica gel, alumina, Florisil® and molecular sieves. The materials are considerably more polar than the organic polymeric adsorbents, leading to the efficient collection of polar materials. Unfortunately, water is also efficiently captured leading to rapid deactivation of the adsorbents. Consequently, these materials are seldom used for sampling trace organic components in air, except in cases where relatively high concentrations of certain polar materials are of concern.

Carbon adsorbents are relatively nonpolar compared to the inorganic adsorbents and hence water adsorption is a less significant problem, although the problem may still prevent analysis in certain applications. The carbon based materials tend to exhibit much stronger adsorption properties than organic polymeric adsorbents, hence allowing efficient collection of volatile materials such as vinyl chloride. However, the strong adsorption on carbon adsorbents can be a disadvantage in cases where recovery by thermal desorption of less volatile materials such as benzene or toluene is desired because of the excessive temperatures required (e.g., 400°C).

There are a variety of carbon based adsorbents available with widely varying adsorption properties. The commonly available classes of carbon adsorbents include:

(1) Various types of conventional activated carbons
(2) Carbon molecular sieves
(3) Carbonaceous polymeric adsorbents

Conventional activated carbons have a microporous structure which leads to difficulty in recovering adsorbed materials and hence this material is rarely used in trace organic sampling.

Adsorption

Adsorption is the phenomenon by which gases, liquids and solutes within liquids are attracted, concentrated and retained at a boundary surface. The boundary surface may be the interface between a gas and liquid, liquid and liquid, gas and solid, liquid and solid, or solid and solid. Of the various boundary surfaces, the adsorption mechanism between liquid and solid, and gas and solid have received the most attention – the former with respect to removal of substances from solution with a solid adsorbent (e.g., purification), and the latter with respect to removing gaseous pollutants on solid adsorbents of high surface area.

Investigation of the adsorption of gases on various solid surfaces has revealed that the operating forces are not the same in all cases. Two types of adsorption have been recognized: (1) physical (or van der Waals), adsorption (physiosorption) and (2) chemical adsorption (chemisorption).

(1) *Physical Adsorption*—In physical adsorption, the attractive forces consist of van der Waals' interactions, dipole-dipole interactions, and/or electrostatic interactions. These forces are similar to those causing the condensation of a gas to a liquid. The process is further characterized by low heats of adsorption, on the order of 2 to 15 kilocalories per mole of adsorbate, and by the fact that adsorption equilibrium is reversible and rapidly established.

Physical adsorption is a commonly occurring process. For example, this is the type of adsorption that occurs when various gases are adsorbed on charcoal. If the temperature[1] is low enough, any gas will be physically adsorbed to a limited extent. The quantity of various gases adsorbed under the same conditions is roughly a function of the ease of condensation of the gases. The higher the boiling point or critical temperature[1] of the gas, the greater is the amount adsorbed. This concept will be discussed in more detail subsequently.

(2) *Chemical Adsorption*—In contrast to physical adsorption, chemical adsorption is characterized by high heats of adsorption, on the order of 20 to 100 kilocalories per mole of adsorbate, which leads to a much stronger binding of the gas molecules to the surface. Heats of adsorption are on the same order of magnitude as chemical reactions, and it is evident that the process involves a combination of gas molecules with the adsorbent to form a surface compound. This type adsorption resembles chemical bonding and is called chemical adsorption, activated adsorption, or chemisorption. For example, in the adsorption of oxygen on tungsten it has been observed that tungsten trioxide distills from the tungsten surface at about 1200 K. However, even at temperatures above 1200 K, oxygen remains on the surface, apparently as tungsten oxide. Additional examples of chemical adsorption are the adsorption of carbon dioxide on tungsten, oxygen on silver, gold on platinum, and carbon and hydrogen on nickel.

A comparison of physical and chemical adsorption can be made by considering the adsorption of oxygen on charcoal. If oxygen is allowed to reach equilibrium with the charcoal at 0°C, most of the oxygen may later be removed from the charcoal by evacuating the system at 0°C with a vacuum pump. However, a small portion of the oxygen cannot be removed from the charcoal no matter how much the pressure is

[1]Critical temperature may be defined as that temperature above which it is impossible to liquefy a gas no matter how high an external pressure is applied.

decreased. If the temperature is now increased, oxygen plus carbon monoxide and carbon dioxide is released from the charcoal. Thus, most of the oxygen is physically adsorbed and can be easily removed, but a small quantity undergoes a chemical reaction with the adsorbent and is not readily removed. In some cases, chemical adsorption may be preceded by physical adsorption, the chemical adsorption occurring after the adsorbent has received the necessary activation energy.

Variables Affecting Gas Adsorption

The quantity of a particular gas that can be adsorbed by a given amount of adsorbent will depend on the following factors: (a) concentration of the gas in the immediate vicinity of the adsorbent; (b) the total surface area of the adsorbent; (c) the temperature of the system; (d) the presence of other molecules competing for a site on the adsorbent; and (e) the characteristics of the adsorbate such as weight, electrical polarity, and chemical reactivity. Ideal physical adsorption of a gas would be favored by a high concentration of material to be adsorbed, a large adsorbing surface, freedom from competing molecules, low temperature, and by aggregation of the adsorbate into a form that conforms with the pore size of the attracting adsorbent.

(1) *Types of Adsorption Isotherms*—The graphic plots of adsorption isotherms yield a wide variety of shapes. Six general types of isotherms have been observed in the adsorption of gases on solids; these are illustrated in Figure 5.1. In physical adsorption, all six

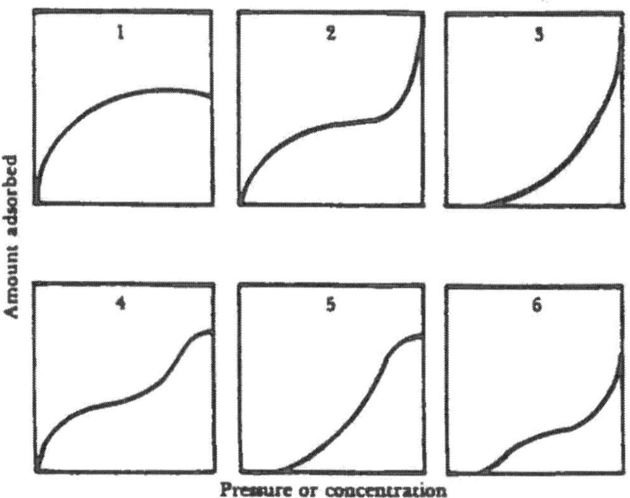

Figure 5.1 Gas adsorption isotherms.

isotherms are encountered, while in chemisorption, only Type 1 occurs.

- Type 1—This type represents the adsorption of a single layer of gas molecules on the adsorbent. There is no interaction between the adsorbed molecules.
- Type 2—This isotherm begins like Type 1 but is modified at high pressure by multilayer adsorption. There is definite interaction between the layers of adsorbed gas molecules.
- Type 3—This type of isotherm is rare. It occurs only when initial adsorption favors a very few strong sites. The interaction between adsorbed molecules is so strong that vacant sites next to occupied sites are stronger than any other vacant sites. In this type of adsorption, the number of effective sites increases with coverage of the adsorbent.
- Types 4 and 5—These two are similar to Types 2 and 3, respectively, except that they continue to exhibit adsorption at high adsorbent coverage.
- Type 6—This type resembles Type 3 with monolayer adsorption first and then continued deposition of a multilayer film.

(2) *Adsorbate Characteristics*—The major adsorbate characteristics affecting the amount of gas adsorbed are the ease of liquefaction of the gas, adsorbate size, concentration of the gas, and presence of other gases.

The specificity by which certain gases are adsorbed on solid adsorbents is illustrated in Table 5.2, where the volumes of different gases adsorbed by one gram of charcoal at 15°C are tabulated.

(3) *Adsorbate Size*—The size of the gas molecule to be removed by

TABLE 5.2. Adsorption of Gases on One Gram of Charcoal at 15°C.*

Gas	Volume Adsorbed (cc)	Critical Temperature (K)
H_2	4.7	33
N_2	8.0	126
CO	9.3	134
CH_4	16.2	190
CO_2	48.0	304
HCl	72.0	324
H_2S	99.0	373
NH_3	181.0	406
$C1_2$	235.0	417
SO_2	380.0	430

*Volumes of gases have been reduced to standard conditions (0°C and 1 atmosphere pressure).

adsorption is characterized by a lower and upper range. The lower size limit is imposed on physical adsorption by the requirement that the pollutant must be higher in molecular weight than the normal components of air. In general, gases with molecular weights greater than 45 are readily removed by physical adsorption. This size includes most odorous and toxic gases of air pollution interest. Gases of interest of lower molecular weight, such as formaldehyde and ammonia, may be removed by chemical adsorption methods using appropriately impregnated adsorbents.

For the upper limit, the individual particles must be sufficiently small so that Brownian motion, or kinetic velocities, will ensure effective contact by collision between them and the granular adsorbent. Although moderate efficiencies may be obtained for very fine mists, the upper limit is generally in the range of molecular size.

(4) *Gas Concentration*—As seen from the examination of adsorption isotherms the quantity of gas adsorbed is a function of the gas concentration or pressure. An increase in concentration or pressure in the vicinity of the adsorbent results in an increase of the total amount of gas adsorbed.

(5) *Presence of Other Gases*—Since the presence of additional gas molecules in a particular adsorbent-adsorbate system causes competition for the limited number of adsorption sites present, the observed effect is a reduction in the amount of adsorbate removed.

(6) *Adsorption Losses in Air Sampling*—Each adsorption medium used in atmospheric sampling has different limitations and problems. The problems most frequently encountered are:

- irreversible adsorption
- variable desorption efficiency
- interference by water vapor

Activated carbon is used extensively because of its high affinity for organic substances. Irreversible adsorption and variable desorption efficiencies are two principal problem areas associated with carbon sampling devices. Carbon can also serve as a potent catalyst creating the possibility of *in-situ* reactions during sampling.

Current Applications of Adsorption in Atmospheric Sampling

Carbon, porous polymers such as Porapack Q, Porapack P, Tenax GC, XAD resins, and polyurethane foam have been used extensively in collecting pesticides, polychlorinated biphenyls and other organic compounds in ambient air. These media can be used in sampling devices, which can be modifications of the hi-volume sampler (see Figure 5.2). Here polyurethane foam is used to collect organics, namely PCBs.

Figure 5.3 illustrates an adsorbent sampling cartridge, and Figure 5.4 shows a cartridge placed in a thermal desorption system.

Carbon molecular sieves have a spherical, macroporous structure which leads to better recovery of adsorbed materials (relative to conventional activated carbons). These materials are sold under various trade names (e.g., Spherocarb, Carbosphere, Carbosieve and Ambersorb XE-347) and have been used to a limited extent for the determination of volatile organics such as vinyl chloride and methylene chloride. Much more work needs to be done in this area before the usefulness of such adsorbents can be established.

Carbonaceous polymeric adsorbents are described as hard, nondusting spheres with properties intermediate between activated carbon and organic polymeric adsorbents. These materials are available from Rohm and Haas Chemical Company under the trade name "Ambersorb" (XE-340, XE-347, XE-348). Of the three materials, Ambersorb XE-347 is classified as a carbon molecular sieve, XE-348 is most similar to activated carbon, and XE-340 is most similar to organic polymeric adsorbents. XE-340 appears to have some attractive features in terms of trace organic sampling and has

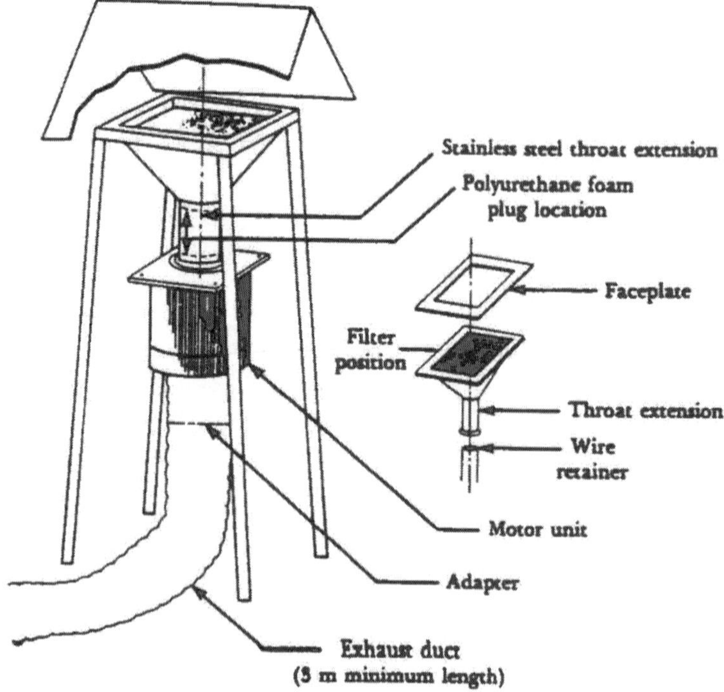

Figure 5.2 Assembled sampler and shelter with exploded view of the filter holder.

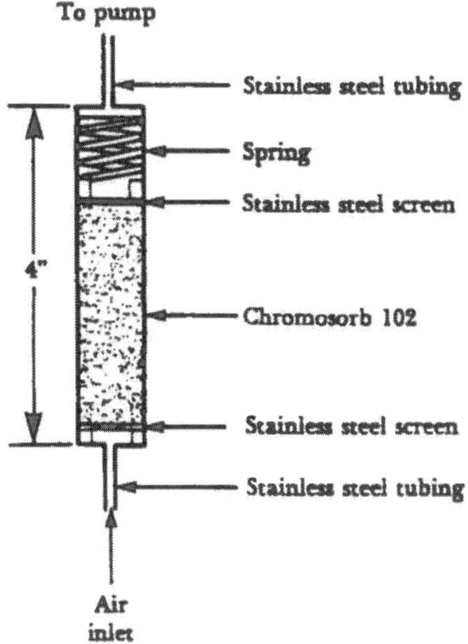

Figure 5.3 High-speed organic vapor collector.

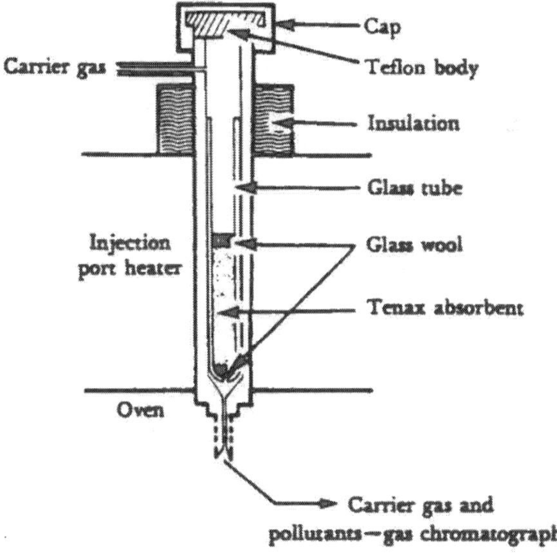

Figure 5.4 Desorption of pollutants from a Tenax GC cartridge.

been shown to be useful for volatile compounds in the C_4 to C_6 boiling point range, a volatility range not covered by organic polymeric adsorbents such as Tenax.

In selecting particular adsorbent materials for sampling trace organics, one must consider both the capture process and the compound recovery process. In general, one of two recovery processes are employed, thermal desorption or solvent extraction. Thermal desorption is most useful for compounds having boiling points less than 300°C, whereas solvent extraction is most useful for compounds boiling above 150°C. Thermal desorption is an attractive approach in many cases, since the entire sample can be introduced onto the analytical instrument compared to solvent extraction where only a small fraction of the extract (e.g., 1–10 percent) can be introduced. Furthermore, the thermal desorption process is more readily automated and does not require disassembly of the sampling cartridge prior to analysis.

Solvent extraction offers the advantage of being able to adjust the concentration of analyte introduced into the analytical system (i.e., to remain within the working range of the instrument) and also allows replicate analysis of a sample, features not available for most thermal desorption systems. Use of solvent extraction also avoids the problem of thermal decomposition of labile compounds during the analysis step.

Although a wide variety of approaches employing solid adsorbents may be feasible for any given monitoring situation, the following summary offers useful guidance for generalized cases:

- Thermal desorption of organic polymeric adsorbents (especially Tenax GC) is useful for compounds boiling between 60–300°C, exclusive of highly polar compounds such as methanol and acetone.
- Solvent extraction of organic polymeric adsorbents (especially XAD-2) is most useful for compounds boiling above 150°C and can in some cases be extended to more volatile compounds, depending on solvent and mode of concentration.
- Thermal desorption of carbon adsorbents (especially carbon molecular sieves or Ambersorb XE-340) may be useful for compounds boiling in the range 0–70°C, including vinyl chloride. However, the high desorption temperatures required (350–400°C) may lead to degradation of certain labile compounds. If medium or high concentrations of compounds are of concern, solvent desorption of carbon adsorbents may be a useful alternative.
- Inorganic adsorbents such a silica gel or alumina are not generally useful for ambient air monitoring because of the water deactivation problem.

Cryogenic Collection

The collection of atmospheric organics by condensation in a cryogenic trap is an attractive alternative to adsorption or impinger collection. The primary advantages of this technique include:

- A wide range of organic materials can be collected.
- Contamination problems with adsorbents and other collection media are avoided.
- The sample is immediately available for analysis without further workup.
- Consistent recoveries are generally obtained.

However, an important limitation of the technique is condensation or large quantities of moisture, and lesser amounts of certain reactive gases (SO_2, NO_x, etc.). The principles of cryogenic sampling are described in Stern's *Air Pollution* series.

Cryogenic sampling can be accomplished in a variety of ways, depending on the desired detection limit and compounds of interest. The important parameters to be specified include:

- choice of cryogen
- trap design
- method of sample recovery
- method of analysis

Useful cryogens include liquid oxygen, liquid argon, dry ice-solvent systems, and ice water. Liquid nitrogen is not an acceptable cryogen, because large quantities of air will condense. Ice water is not cold enough for collecting organics in ambient air, except for relatively nonvolatile compounds. Dry ice-solvent systems should be employed with caution because of the high probability of contaminating the sample with relatively large quantities of solvent. Liquid oxygen or argon appear to be the most generally applicable cryogens, although the safety hazards associated with liquid oxygen make its use less attractive.

The design of a suitable sample trap is extremely important in cryogenic sampling. One must ensure that the air residence time in the trap during sampling is great enough to allow cooling of the gas stream and condensation of the analyte of interest. The trap material must be able to withstand the wide temperature range involved in the sample process. The trap design must also be appropriate for the sample recovery step, allowing efficient recovery without loss or contamination of the sample. For highly volatile materials, the inclusion of an adsorption medium such as silica gel may be necessary to obtain good collection efficiency.

Cryogenically collected samples can be recovered either by flash evaporation into an analytical instrument or by solvent flashing of the trap. The former approach is preferable in most cases, since it allows more sensitive detection and avoids contamination (e.g., from solvents). However, the solvent flushing approach can be accomplished with less elaborate equipment in the field and may be preferable for the analysis of high concentrations of material.

A generally applicable cryogenic sampling approach for monitoring trace organics in ambient air has been described. This system involves the use of a small (3mm ID) freeze out trap held at liquid argon or liquid oxygen temperature with packed with silanized glass beads. A 50–500 ml volume of ambient air is drawn through the trap using an evacuated tank/manometer assembly as the pumping and volume measurement system. The trap is directly connected to a six-part stainless steel valve and following sampling collection, the condensed material is flash evaporated onto a capillary GC/FID system for analysis. Detection limits on the order of 0.5 to 1 ppbv can be achieved using this approach. Even lower detection limits can be achieved for certain compounds with the use of selective detectors, such as electron capture (ECD).

This technique requires transport of the analytical system to the monitoring site, or collection of a whole air sample in an evacuated cylinder with subsequent laboratory analysis. For higher boiling compounds adsorption onto container walls may represent a significant problem if the latter approach is employed.

Impinger Collection

Impinger or "bubbler" collection involves passing the gas stream through an organic solvent or other suitable liquid and capture of the analyte by partioning into the solvent (special derivatization impinger systems are discussed later in this document). This technique is not generally applicable for trace organic analysis because large volumes of air cannot be sampled, due to solvent evaporation during the sampling process. However, this technique is relatively simple and may be useful in high level (ppm) monitoring situations.

The impinger should be designed so that contact between the air and solvent is maxmized, either by the use of fritted diffusers or capillary jets. Some typical impinger designs are given in Stern's book. In general, the impinger systems should be cooled in ice water so as to reduce solvent loss during sampling.

Whole Air Collection

Collection of whole air samples using evacuated glass bulbs, stainless steel canisters, or similar devices is probably the simplest sampling approach, and can be useful in many situations. An obvious limitation of this approach is that the sample components of interest may be adsorbed or decomposed through interaction with the container walls. At very high analyte levels (e.g., several ppm), condensation may be a problem. Consequently, this approach is most useful for relatively stable, volatile compounds such as hydrocarbons and chlorinated hydrocarbons with boiling points less than 150°C. However, certain compounds within these classes pose storage stability problems (e.g., carbon tetrachloride interacts with stainless steel surfaces and is lost). Preconditioning surfaces (e.g., formation of an oxide coating) or selection of alternate container materials can circumvent these problems in many cases. In all cases, the container must be flushed (with moderate heating if possible) with zero grade nitrogen or air prior to sampling in order to remove trace contaminants.

Containers used for whole air sampling can be roughly categorized as rigid or nonrigid devices. Nonrigid devices include Tedlar® or Teflon® plastic bags. Generally, such devices are used to collect samples for analysis within a few hours, since the rate of leakage and/or permeation of materials into and out of the bag is relatively high.

Rigid containers have an advantage in that leakage and/or permeation rates are generally low and samplers can be stored for several days or even several months for certain compounds. Examples of rigid containers include:

- glass bulbs
- gas-tight syringes
- stainless steel cylinders or canisters

Sampling into such devices can be accomplished either by evacuating the container in advance and then allowing sample to enter the container or by having both inlet and outlet valves on the container and pumping sample through the container until equilibrium is obtained (e.g., after 5–10 container volumes of sample have been flushed through the system). The first approach has the advantage that no sampling pump is required in the field. However, the latter approach has the advantage that equilibrium with the container walls is more readily achieved and high vacuum seals are not required.

An alternate sampling approach is to pressurize a whole air sample into a stainless steel cylinder. This approach has proven useful in ambient air monitoring studies wherein electropolished 6-liter stainless steel cans were pressurized with ambient air to approximately 15 psi. Under these conditions, relatively volatile materials having boiling points less than 120°C had stable concentrations for one to two weeks. This approach has the advantage that a relatively large volume of air can be collected and transported to the laboratory for analysis. However, condensation of material will become a significant problem as the container pressure is increased.

Derivatization Techniques

A fundamental limitation of the various sampling techniques discussed so far is the decomposition of reactive compound during collection or transport of the samples. Two approaches can be used to circumvent this problem. One obvious approach is to use a direct analysis technique in which the analyte concentration is determined without physically isolating the air sample (e.g., total hydrocarbon analysis).

An alternate approach, discussed in this section, is to stabilize reactive compounds by combining them with a derivatizing reagent during the sampling process. In many cases, derivatizing reagents can be chosen which not only stabilize the compound but enhance its detectability. Such schemes are available for determining aldehydes, phosgene and certain other reactive compounds.

In certain cases, such as the 4-aminoantipyrene method for phenols and the ninhydrin method for amines the derivatization step is used solely to enhance detectability and may be done in the laboratory, rather than the field. In this section, only derivatization techniques for field use are considered. Derivatization schemes for specific compounds or compound classes are discussed later.

Derivatization reagents for field use can be held either in an impinger or on a solid absorbent. The impinger approach is most convenient, since reagents can be prepared and stored as liquid solutions whereas solid adsorbent systems require more elaborate preparation and storage procedures. However, the solid adsorbent approach is more sensitive in many cases, because larger volumes of air can be sampled. A study of both these approaches for aldehyde derivatization has been reported. In this particular example, the impinger approach appears most useful, since it provides adequate sensitivity and is less susceptible to humidity effects on analyte recovery.

CHAPTER 6

Odorous Materials

TESTING OLFACTORY SENSITIVITY

INTRODUCTION

THIS information relates to the testing of the olfactory sensitivity of individuals. This is a relatively simple and straightforward technique that requires the use of squeeze bottles containing various concentrations of a specific odorant (such as Thiophane or Carbinol). The prepared chemicals can be obtained from John Amoore, Olfacto Labs, P.O. Box 757, El Cerrito, CA 94530.

This technique uses:

- lightweight, miniature squeeze bottles
- nonspill design with inert absorbent
- pleasant-smelling test odorant
- labeled, sectional storage case
- combined score-sheet and instructions
- logical "decismel scale" of odor thresholds (modeled on the familiar decibel scale for hearing)

Squeeze bottles with flip-top caps deliver puffs of odorized air to the nose. A graded series of odor concentrations (with paired blank bottles, if desired) enables the person's olfactory threshold to be accurately measured in a few minutes.

Historically, this procedure has been used in clinics to help in the evaluation, diagnosis, recovery of persons with nasal problems. Abnormally low smell sensitivity (hyposmia or anosmia) can be indicative of intranasal polyps or carcinoma, atrophic rhinitis, tumor or abscess near the anterior fossa, and hypogonadotrophic hypogonadism. Temporary or chronic loss of smell sensitivity should be evaluated and monitored in head injury, occupational exposure to chemicals, and radiation therapy to the head.

Clinics specializing otolaryngology, endocrinology, allergy, neurology, radiation therapy, or occupational medicine most often need quantitative olfactory testing, using the closely spaced odor intensities of Kit No. 11. For saving medical office time, the wider intervals of the utility Kit No. 12 are effective, though with some reduction of precision. The needs of many primary-care physicians are met by screening Kits No. 13 or 14, which omit the paired blank bottles, and are directed to the most significant odor intensity levels. For medical schools and population studies, educational Kit No. 15 is recommended.

The flowery-smelling odorant is "PM-Carbinol" (phenylethyl methyl ethyl carbinol) diluted in mineral oil. Each kit has a 1-year service life and is capable of testing over 100 patients.

(1) *Research Kit No. 11*—Eighteen levels of odor intensity, covering the full normal and hyposmia ranges (every 5 decismels from −30 to +55 decismels). Each odor bottles has its matched blank (odorless) bottle for exact discriminations.

(2) *Utility Kit No. 12*—Nine levels of odor intensity covering the same range (every 10 decismels).

(3) *Scaling Kit No. 13*—Seven intensity levels, every 10 decismels from −5 through +55 decismels (dS). One blank bottle is provided for cross-checking.

(4) *Screening Kit No. 14*—Three key levels of intensity (−5, 25, 55 dS) for quickly defining normality, hyposmia and anosmia, with one blank.

(5) *Educational Kit No. 15*—Nine levels in the normal range, every 5 dS fro −20 to +20, with matched blanks, for use in medical school.

Recommended Pre-Test Procedure

General

Avoid using cosmetics, perfumes, after-shave or scented soap, and for 15 minutes prior to testing avoid smoking, food, drink candy or chewing gum. Have the test subjects follow the same precautions (this will aid in the accuracy of the test). The test administrator and test subject should wash their hands prior to the start of each test.

Each pair of bottles must be conditioned to the current test environment before being given to the test subject. Flip open the caps and squeeze each bottle twice. (Air from a different source may be in each bottle; by exchanging the head-space air with the current environment an increased accuracy will result.)

Explain the Blank

Tell the subject that a selection will be made between a pair of bottles; one bottle having an odor and the other a blank. In absolute terms, nothing can be defined as odor-free; the blank bottle has been processed to be as odor-free as possible. This may be different from the environment in which the test is being administered. The subject should be directed to select between the bottles by evaluating odor intensity or difference.

Demonstrating the Test

Select a pair of bottles from near the middle of the normal range for adults (in this test +5 decismels; 5 dS). Demonstrate on yourself; flip open the bottle caps, hold the orifice of one bottle about an inch from the nostril and squeeze the bottle several times while sniffing, then try the same bottle on the other nostril (the nose tends to breathe through one nostril preferentially for a time and then switches over). Repeat the procedure with the other bottle.

Hand these bottles to the subject to practice the test procedure. After selecting a bottle for having an odor or a stronger odor, the subject should hand you that bottle; check the label on the base of the bottle. If the choice is wrong, or if the subject says no odor is distinguishable, offer progressively higher decismels bottles (in 10 dS steps) until an odor can confidently be detected. Repeated sniffs are permitted, before making the choice. Detection only is required, not recognition of the odor.

Beginning the Test

Once the subject feels comfortable with the test, take the bottles back and rotate or randomize both bottles out of sight of the subject. It is preferable to continue the randomization until the tester has forgotten which is the odorous bottle (doubled-blind test). Hand the two bottles back to the subject, to select the bottle having the stronger odor. Enter the result at the appropriate decismel level in the 1st column of the score sheet, "C" for correct and "X" for wrong. (From now on, do not comment on the selections, as this may influence the result of the test.) Take the second bottle back and randomize both bottles out of sight of the subject; hand them back and repeat the procedure a second and a third time. Enter the results in the 2nd and 3rd columns of the score sheet. In order for the subject to pass a given decismel level, the correct selection must be made all three times (no errors permitted).

When done, close the caps, put the bottles back in the box and select the next lower decismel pair. Hand these bottles to the subject and repeat the test procedure three times, entering the results in the appropriate spaces of the score sheet.

If the subject again identifies the correct bottle all three times, repeat the procedure at the next lower dS level. Continue in this way until the subject makes a wrong selection. Note that the subject has to make a choice (forced choice); a mere statement that he or she cannot detect any difference is not acceptable.

The subject's olfactory threshold for this substance, on this occasion, is the lowest decismel level at which all three selections were correct. At least two consecutive decismel levels should be passed, in order to establish a valid threshold. If only one level was passed before an error was made, then the subject must be tested on, and pass, the next higher dS level to verify the test result.

Concluding the Odor Threshold Test

Note the olfactory threshold in decismels and the corresponding evaluation at the foot of the score sheet. For greater precision on any one subject, the compete procedure may be repeated. Other odors may be tested in the same way, if additional smell test kits are available.

Test Procedure

Start with the pair of bottles for the 5 decismels level. Press to open the disc-tops, and squeeze the bottle to direct the odorized air closely toward the subject's nostrils, while he or she is sniffing. (Capable subjects may handle the bottles themselves.) Ask the subject to indicate which bottle has the odor (or stronger odor). Check label on base of bottle, and enter his selection (C = correct, X = wrong) in 1st column on Table 6.1.

If the choice is correct, randomize the positions of the bottles (out of sight of the subject) and repeat procedure for the 2nd and 3rd tests. Continue with successively lower decismel levels, to determine the subject's olfactory threshold (that is, the lowest decismels with all three correct selections).

If the initial test with 5 decismels is wrong, or the subject says he cannot detect any odor difference, offer increasing decismels until the odor is detected, then proceed as above.

Each test odor bottle should always be compared with its own matched blank bottle. Close all disc-tops after use, and replace the bottles in the box. Enter the subject's olfactory threshold in decismels on the bottom line, and the corresponding evaluation from the guide at the right side of Table 6.1.

TABLE 6.1. Olfactory Threshold Test.

(PM-CARBINOL)	DECI-SMELS LEVEL	TESTS 1st	2nd	3rd	AN-OSMIA
<u>Questionnaire</u> **Name:**___________________ **Age:**____ **Tester:**________________ **Date:**_______ To avoid contaminating the bottles, and for an accurate test result, it is essential that: Subject has not smoked in last 15 min.? No food, drink or candy in last 15 min.? No odor or perfume on the hands or face? No signs of a nasal infection or allergy? Subject doesn't work near perfumed goods? No odors in the area used for this test?	55				
	50				H Y P O S M I A
	45				
	40				
	35				
	30				
	25				
	20				
	15				N O R M A L
	10				
	5				
	0				
	-5				R A N G E
	-10				
	-15				
	-20				
	-25				
	-30				HYPER OSMIA

<u>Olfactory evaluation</u>
PM-Carbinol threshold is ____ decismels.

Interpretation of the Result

This scale of olfactory sensitivity is based on research data obtained with 100 healthy subjects. The concentration of PM-Carbinol at the reference point (zero decismel) is adjusted to match the average threshold of the group (normalized to age 20).

The normal range of olfactory thresholds (95% of subjects ages 20

TABLE 6.2. Basis of the Decismel Scale.

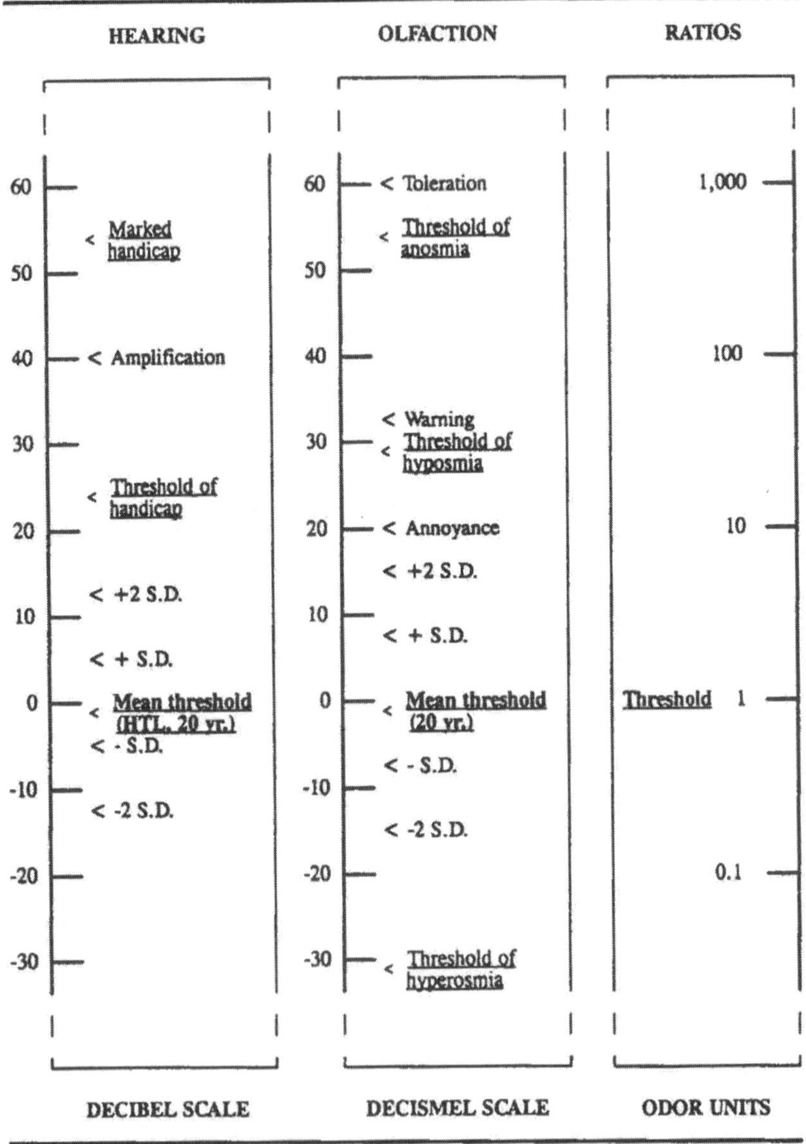

through 70 is from −25 through +25 decismels, with standard deviation ±12 decismels. The reproducibility of the test for a given subject is ±5 decismels.

From ages 10 through 70, there is an average increase in threshold of 0.3 decismel per year. After adjustment for age, there are negligible differences between the genders, or between (moderate) smokers and nonsmokers.

Olfactory Abnormalities

Any subject who cannot detect 55 decismels is considered to have virtually no olfactory (1st cranial) nerve function. This condition, known as (general) anosmia, affects about 0.2% of the population. It has been observed in a variety of pathological conditions, and may also occur spontaneously. (Such subjects may be able to detect ammonia and other irritants by sensations from the trigeminal nerve.)

Thresholds in the range of 30 through 55 decismels represent hyposmia, which is observed in roughly 5% of the population. This is often, though not necessarily, associated with clinical abnormalities.

Ability to discriminate the odor at −30 decismels or below indicates hyperosmia, in about 1% of the population. This may be inconsequential, but could be indicative of a clinical problem.

Technical Notes

PM-Carbinol is phenylethyl-methyl-ethyl-carbinol (an FDA-approved flavoring) diluted in White Oil USP, which is adsorbed by an inert matrix. It has a pleasant flowery odor. Detection only is required of the subject, not odor recognition.

The decismel scale of odor levels is closely modeled on the decibel scale of sound levels. By definition:

$$\text{Odor level in decismels} = 20 \times \log_{10}\left(\frac{\text{vapor concentration}}{\text{reference concentration}}\right)$$

This is a logarithmic scale which yields a normal distribution of olfactory thresholds for the majority of the population. The vapor concentration doubles every 6 decismels, and the perceived odor intensity doubles every 10 decismels (see Table 6.2).

The preferred room temperature for storage of the kit and for conducting an accurate test is 63−77°F (16−25°C). Do not store below 40°F or above 90°F (4−32°C). Keep the bottles upright in storage or transport. After 12 months or 100 subjects, whichever occurs sooner, the set of bottles should be replaced.

Reference Method for the Determination of Suspended Particulate Matter in the Atmosphere (High-Volume Method)

1.0 *Applicability.*

1.1 This method provides a measurement of the mass concentration of total suspended particulate matter (TSP) in ambient air for determining compliance with the primary and secondary national ambient air quality standards for particulate matter as specified in § 50.6 and § 50.7 of this chapter. The measurement process is nondestructive, and the size of the sample collected is usually adequate for subsequent chemical analysis. Quality assurance procedures and guidance are provided in Part 58, Appendices A and B, of this chapter and in References [1] and [2].

2.0 *Principle.*

2.1 An air sampler, properly located at the measurement site, draws a measured quantity of ambient air into a covered housing and through a filter during a 24-hr (nominal) sampling period. The sampler flow rate and the geometry of the shelter favor the collection of particles up to 25–50 μm (aerodynamic diameter), depending on wind speed and direction [3]. The filters used are specified to have a minimum collection efficiency of 99 percent for 0.3 μm (DOP) particles (see Section 7.1.4.).

2.2 The filter is weighed (after moisture equilibration) before and after use to determine the net weight (mass) gain. The total volume of air sampled, corrected to EPA standard conditions (25°C, 760 mm Hg [101 kPa]), is determined from the measured flow rate and the sampling time. The concentration of total suspended particulate matter in the ambient air is computed as the mass of collected particles divided by the volume of air sampled, corrected to standard conditions, and is expressed in micrograms per standard cubic meter ($\mu/stdm^3$). For samples collected at temperatures and pressures significantly different than standard conditions, these corrected concentrations may differ substantially from actual concentrations (micrograms per actual cubic meter), particularly at high elevations. The actual particulate matter concentration can be calculated from the corrected concentration using the actual temperature and pressure during the sampling period.

3.0 *Range*.

3.1 The approximate concentration range of the method is 2 to 750 μg/std m^3. The upper limit is determined by the point at which the sampler can no longer maintain the specified flow rate due to the increased pressure drop of the loaded filter. This point is affected by particle size distribution, moisture content of the collected particles, and variability from filter to filter, among other things. The lower limit is determined by the sensitivity of the balance (see Section 7.10) and by inherent sources of error (see Section 6).

3.2 At wind speeds between 1.3 and 4.5 m/sec (3 and 10 mph), the high volume air sampler has been found to collect particles up to 25 to 50 mm, depending on wind speed and direction [3]. For the filter specified in Section 7.1, there is effectively no lower limit on the particle size collected.

4.0 *Precision.*

4.1 Based upon collaborative testing, the relative standard deviation (coefficient of variation) for single analyst precision (repeatability) of the method is 3.0 percent. The corresponding value for interlaboratory precision (reproducibility) is 3.7 percent [4].

5.0 *Accuracy.*

The absolute accuracy of the method is undefined because of the complex nature of atmospheric particulate matter and the difficulty in determining the "true" particulate matter concentration. This method provides a measure of particulate matter concentration suitable for the purpose specified under Section 1.0, Applicability.

6.0 *Inherent Sources of Error.*

6.1 *Airflow variation.* The weight of material collected on the filter represents the (integrated) sum of the product of the instantaneous flow rate times the instantaneous particle concentration. Therefore, dividing this weight by the average flow rate over the sampling period yields the true particulate matter concentration only when the flow rate is constant over the period. The error resulting from a nonconstant flow rate depends on the magnitude of the instantaneous changes in the flow rate and in the particulate matter concentration. Normally, such errors are not large, but they can be greatly reduced by equipping the sampler with an automatic flow controlling mechanism that maintains constant flow during the sampling period. Use of a constant flow controller is recommended.[2]

6.2 *Air volume measurement.* If the flow rate changes substantially or nonuniformly during the sampling period, appreciable error in the estimated air volume may result from using the average of the presampling and

[2]At elevated altitudes, the effectiveness of automatic flow controllers may be reduced because of a reduction in the maximum sampler flow.

postsampling flow rates. Greater air volume measurement accuracy may be achieved by (1) equipping the sampler with a flow controlling mechanism that maintains constant air flow during the sampling period,[3] (2) using a calibrated, continuous flow rate recording device to record the actual flow rate during the sampling period and integrating the flow rate over the period, or (3) any other means that will accurately measure the total air volume sampled during the sampling period. Use of a continuous flow recorder is recommended, particularly if the sampler is not equipped with a constant flow controller.

6.3 *Loss of volatiles.* Volatile particles collected on the filter may be lost during subsequent sampling or during shipment and/or storage of the filter prior to the postsampling weighing [5]. Although such losses are largely unavoidable, the filter should be reweighed as soon after sampling as practical.

6.4 *Artifact particulate matter.* Artifact particulate matter can be formed on the surface of alkaline glass fiber filters by oxidation of acid gases in the sample air, resulting in a higher than true TSP determination [6,7]. This effect usually occurs early in the sample period and is a function of the filter pH and the presence of acid gases. It is generally believed to account for only a small percentage of the filter weight gain, but the effect may become more significant where relatively small particulate weights are collected.

6.5 *Humidity.* Glass fiber filters are comparatively insensitive to changes in relative humidity, but collected particulate matter can be hygroscopic [8]. The moisture conditioning procedure minimizes but may not completely eliminate error due to moisture.

6.6 *Filter handling.* Careful handling of the filter between the presampling and postsampling weighings is necessary to avoid errors due to loss of fibers or particles from the filter. A filter paper cartridge or cassette used to protect the filter can minimize handling errors (see Reference [2], Section 2).

6.7 *Nonsampled particulate matter.* Particulate matter may be deposited on the filter by wind during periods when the sampler is inoperative [9]. It is recommended that errors from this source be minimized by an automatic mechanical device that keeps the filter covered during nonsampling periods, or by timely installation and retrieval of filters to minimize the nonsampling periods prior to and following operation.

6.8 *Timing errors.* Samplers are normally controlled by clock timers set to start and stop the sampler at midnight. Errors in the nominal 1,440-min sampling period may result from a power interruption during the sampling period or from a discrepancy between the start or stop time recorded on the

[3]See footnote 2.

filter information record and the actual start or stop time of the sampler. Such discrepancies may be caused by (1) poor resolution of the timer set-points, (2) timer error due to power interruption, (3) missetting of the timer, or (4) timer malfunction. In general, digital electronic timers have much better set-point resolution than mechanical timers, but require a battery backup system to maintain continuity of operation after a power interruption. A continuous flow recorder or elapsed time meter provides an indication of the sampler run-time, as well as indication of any power interruption during the sampling period and is therefore recommended.

6.9 *Recirculation of sampler exhaust.* Under stagnant wind conditions, sampler exhaust air can be resampled. This effect does not appear to affect the TSP measurement substantially, but may result in increased carbon and copper in the collected sample [10]. This problem can be reduced by ducting the exhaust air well away, preferably downwind, from the sampler.

7.0 *Apparatus.*

(See References [1] and [2] for quality assurance information.)

Note: Samplers purchased prior to the effective date of this amendment are not subject to specifications preceded by *[4].

7.1 *Filter.* (Filters supplied by the Environmental Protection Agency can be assumed to meet the following criteria. Additional specifications are required if the sample is to be analyzed chemically.)

7.1.1 *Size:* 20.3 ± 0.2 × 25.4 ± 0.2 cm (nominal 8 × 10 in).

7.1.2 *Nominal exposed area:* 406.5 cm^2 (63 in^2).

7.1.3 *Material:* Glass fiber or other relatively inert, nonhygroscopic material [8].

7.1.4 *Collection efficiency:* 99 percent minimum as measured by the DOP test (ASTM-2986) for particles of 0.3 μm diameter.

7.1.5 *Recommended pressure drop range:* 42−54 mm Hg (5.6−7.2 kPa) at a flow rate of 1.5 std m^3/min through the nominal exposed area.

7.1.6 *pH:* 6 to 10 [11].

7.1.7 *Integrity:* 2.4 mg maximum weight loss [11].

7.1.8 *Pinholes:* None.

7.1.9 *Tear strength:* 500 g minimum for 20 mm wide strip cut from filter in weakest dimension (see ASTM Test D828-60).

7.1.10 *Brittleness:* No cracks or material separations after single lengthwise crease.

7.2 *Sampler.* The air sampler shall provide means for drawing the air sample, via reduced pressure, through the filter at a uniform face velocity.

7.2.1 The sampler shall have suitable means to:

a. Hold and seal the filter to the sampler housing.

b. Allow the filter to be changed conveniently.

[4] See note at beginning of Section 7 of this Appendix.

c. Preclude leaks that would cause error in the measurement of the air volume passing through the filter.

d. *Manually adjust the flow rate to accommodate variations in filter pressure drop and site line voltage and altitude.[5] The adjustment may be accomplished by an automatic flow controller or by a manual flow adjustment device. Any manual adjustment device must be designed with positive detents or other means to avoid unintentional changes in the setting.

7.2.2 *Minimum sample flow rate, heavily loaded filter:* 1.1 m^3/min (39 ft^3/min).[6]

7.2.3 *Maximum sample flow rate, clean filter:* 1.7 m^3/min (60 ft^3/min).[7]

7.2.4 *Blower motor:* The motor must be capable of continuous operation for 24-hour periods.

7.3 *Sampler shelter.*

7.3.1 The sampler shelter shall:

a. Maintain the filter in a horizontal position at least 1 m above the sampler supporting surface so that sample air is drawn downward through the filter.

b. Be rectangular in shape with a gabled roof, similar to the design shown in Figure A.1.

c. Cover and protect the filter and sampler from precipitation and other weather.

d. Discharge exhaust air at least 40 cm from the sample air inlet.

e. Be designed to minimize the collection of dust from the supporting surface by incorporating a baffle between the exhaust outlet and the supporting surface.

7.3.2 The sampler cover or roof shall overhang the sampler housing somewhat, as shown in Figure A.1, and shall be mounted so as to form an air inlet gap between the cover and the sampler housing walls. *This sample air inlet should be approximately uniform on all sides of the sampler.[8] *The area of the sample air inlet must be sized to provide an effective particle capture air velocity of between 20 and 35 cm/sec at the recommended operational flow rate.[9] The capture velocity is the sample air flow rate divided by the inlet area measured in a horizontal plane at the lower edge of the cover. Ideally, the inlet area and operational flow rate should be selected to obtain a capture air velocity of 25 ± 2 cm/sec.

[5]See footnote 4.

[6]These specifications are in actual air volume units; to convert to EPA standard air volume units, multiply the specifications by (P_b/P_{std}) $(298/T)$ where P_b and T are the barometric pressure in mm Hg (or kPa) and the temperature in K at the sampler, and P_{std} is 760 mm Hg (or 101 kPa).

[7]See footnote 6.

[8]See footnote 4.

[9]See footnote 4.

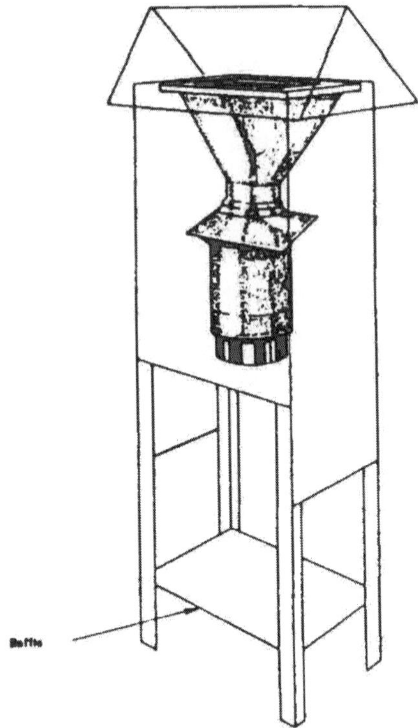

Figure A.1 High-volume sampler in shelter.

7.4. *Flow rate measurement devices.*

7.4.1 The sampler shall incorporate a flow rate measurement device capable of indicating the total sampler flow rate. Two common types of flow indicators covered in the calibration procedure are (1) an electronic mass flowmeter and (2) an orifice or orifices located in the sample air stream together with a suitable pressure indicator such as a manometer, or aneroid pressure gauge. A pressure recorder may be used with an orifice to provide a continuous record of the flow. Other types of flow indicators (including rotameter) having comparable precision and accuracy are also acceptable.

7.4.2 *The flow rate measurement device must be capable of being calibrated and read in units corresponding to a flow rate which is readable to the nearest 0.02 std m^3/min over the range 1.0 to 1.8 std m^3/min.[10]

7.5 *Thermometer*, to indicate the approximate air temperature at the flow rate measurement orifice, when temperature corrections are used.

7.5.1 *Range:* −40° to +50°C (223−323 K).

7.5.2 *Resolution:* 2°C (2 K).

[10]See footnote 4.

7.6 ***Barometer***, to indicate barometric pressure at the flow rate measurement orifice, when pressure corrections are used.

7.6.1 *Range:* 500 to 800 mm Hg (66 – 106 kPa).

7.6.2 *Resolution:* ±5 mm Hg (0.67 kPa).

7.7 *Timing/control device.*

7.7.1 The timing device must be capable of starting and stopping the sampler to obtain an elapsed run-time of 24 hr ± 1 hr (1,440 ± 60 min).

7.7.2 *Accuracy of time setting:* ±30 min, or better (See Section 6.8).

7.8 *Flow rate transfer standard*, traceable to a primary standard (see Section 9.2).

7.8.1 *Approximate range:* 1.0 to 1.8 m^3/min.

7.8.2 *Resolution:* 0.02 m^3/min.

7.8.3 *Reproducibility:* ±2 percent (2 times coefficient of variation) over normal ranges of ambient temperature and pressure for the stated flow rate range (see Reference [2], Section 2).

7.8.4 *Maximum pressure drop at 1.7 std m^3/min;* 50 cm H_2O (5 kPa).

7.8.5 The flow rate transfer standard must connect without leaks to the inlet of the sampler and measure the flow rate of the total air sample.

7.8.6 The flow rate transfer standard must include a means to vary the sampler flow rate over the range of 1.0 to 1.8 m^3/min (35 – 64 ft^3/min) by introducing various levels of flow resistance between the sampler and the transfer standard inlet.

7.8.7 The conventional type of flow transfer standard consists of: An orifice unit with adapter that connects to the inlet of the sampler, a manometer or other device to measure orifice pressure drop, a means to vary the flow through the sampler unit, a thermometer to measure the ambient temperature, and a barometer to measure ambient pressure. Two such devices are shown in Figures A.2(a) and A.2(b). Figure A.2(a) shows multiple fixed resistance plates, which necessitate disassembly of the unit each time the flow resistance is changed. A preferable design, illustrated in Figure A.2(b), has a variable flow restriction that can be adjusted externally without disassembly of the unit. Use of a conventional, orifice-type transfer standard is assumed in the calibration procedure (Section 9). However, the use of other types of transfer standards meeting the above specifications, such as the one shown in Figure A.2(c), may be approved; see the note following Section 9.1.

7.9 *Filter conditioning environment.*

7.9.1 *Controlled temperature:* between 15° and 30°C with less than ±3°C variation during equilibration period.

7.9.2 *Controlled humidity:* Less than 50 percent relative humidity, constant within ±5 percent.

7.10 *Analytical balance.*

7.10.1 *Sensitivity:* 0.1 mg.

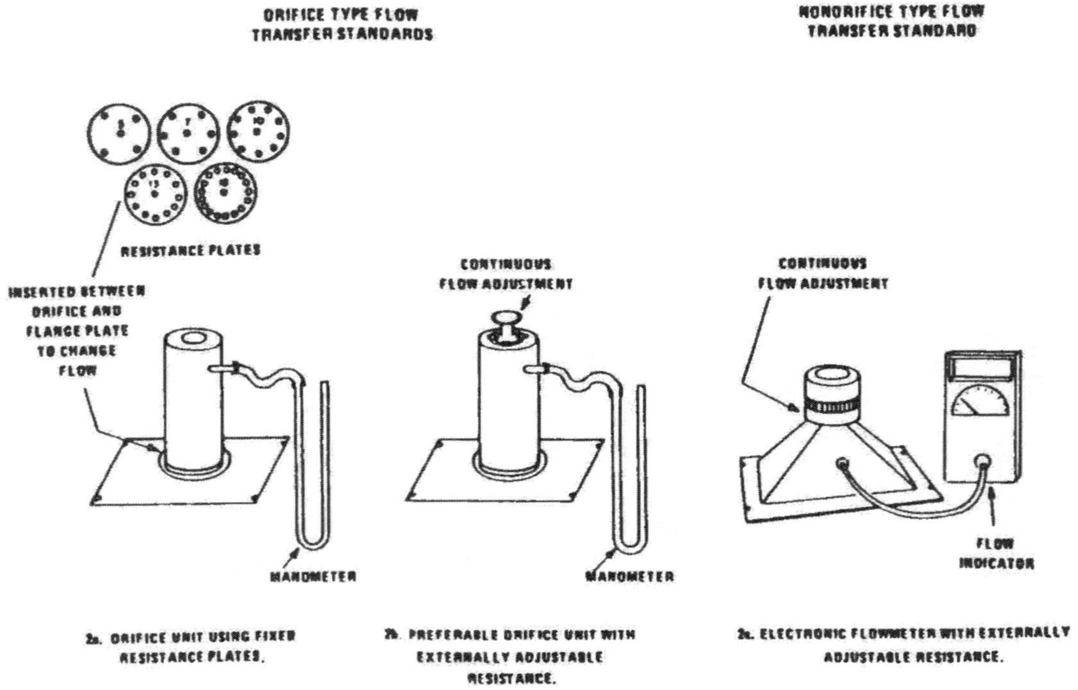

Figure A.2 Various types of flow transfer standards. Note that all devices are designed to mount to the filter inlet area of the sampler.

7.10.2 Weighing chamber designed to accept an unfolded 20.3 × 25.4 cm (8 × 10 in) filter.

7.11 *Area light source,* similar to X-ray film viewer, to backlight filters for visual inspection.

7.12 *Numbering device,* capable of printing identification numbers on the filters before they are placed in the filter conditioning environment, if not numbered by the supplier.

8.0 *Procedure.*

(See References [1] and [2] for quality assurance information.)

8.1 Number each filter, if not already numbered, near its edge with a unique identification number.

8.2 Backlight each filter and inspect for pinholes, particles, and other imperfections; filters with visible imperfections must not be used.

8.3 Equilibrate each filter in the conditioning environment for at least 24-hr.

8.4 Following equilibration, weight each filter to the nearest milligram and record this tare weight (W_i) with the filter identification number.

8.5 Do not bend or fold the filter before collection of the sample.

8.6 Open the shelter and install a numbered, preweighed filter in the sampler, following the sampler manufacturer's instructions. During inclement weather, precautions must be taken while changing filters to prevent damage to the clean filter and loss of sample from or damage to the exposed filter. Filter cassettes that can be loaded and unloaded in the laboratory may be used to minimize this problem (see Section 6.6).

8.7 Close the shelter and run the sampler for at least 5 min to establish run-temperature conditions.

8.8 Record the flow indicator reading and, if needed, the barometric pressure (P_3) and the ambient temperature (T_3) (see note following step 8.12). Stop the sampler. Determine the sampler flow rate (see Section 10.1); if it is outside the acceptable range (1.1 to 1.7 m^3/min [39–60 ft^3/min]), use a different filter, or adjust the sampler flow rate. Warning: Substantial flow adjustments may affect the calibration of the orifice-type flow indicators and may necessitate recalibration.

8.9 Record the sampler identification information (filter number, site location or identification number, sample date, and starting time).

8.10 Set the timer to start and stop the sampler such that the sampler runs 24-hrs, from midnight to midnight (local time).

8.11 As soon as practical following the sampling period, run the sampler for at least 5 min to again establish run-temperature conditions.

8.12 Record the flow indicator reading and, if needed, the barometric pressure (P_3) and the ambient temperature (T_3).

Note: No onsite pressure or temperature measurements are necessary if the sampler flow indicator does not require pressure or temperature corrections (e.g., a mass flowmeter or if average barometric pressure and seasonal average temperature for the site are incorporated into the sampler calibration (see step 9.3.9). For individual pressure and temperature corrections, the ambient pressure and temperature can be obtained by onsite measurements or from a nearby weather station. Barometric pressure readings obtained from airports must be station pressure, not corrected to sea level, and may need to be corrected for differences in elevation between the sampler site and the airport. For samplers having flow recorders but not constant flow controllers, the average temperature and pressure at the site *during the sampling period* should be estimated from weather bureau or other available data.

8.13 Stop the sampler and carefully remove the filter, following the sampler manufacturer's instructions. Touch only the outer edges of the filter. See the precautions in step 8.6.

8.14 Fold the filter in half lengthwise so that only surfaces with collected particulate matter are in contact and place it in the filter holder (glassine envelope or manilla folder).

8.15 Record the ending time or elapsed time on the filter information record, either from the stop set-point time, from an elapsed time indicator, or from a continuous flow record. The sample period must be 1,440 ± 60 min for a valid sample.

8.16 Record on the filter information record any other factors, such as meteorological conditions, construction activity, fires or dust storms, etc., that might be pertinent to the measurement. If the sample is known to be defective, void it at this time.

8.17 Equilibrate the exposed filter in the conditioning environment for at least 24-hrs.

8.18 Immediately after equilibration, reweigh the filter to the nearest milligram and record the gross weight with the filter identification number. See Section 10 for TSP concentration calculations.

9.0 *Calibration.*

9.1 Calibration of the high volume sampler's flow indicating or control device is necessary to establish traceability of the field measurement to a primary standard via a flow rate transfer standard. Figure A.3(a) illustrates the certification of the flow rate transfer standard and Figure A.3(b) illustrates its use in calibrating a sampler flow indicator. Determination of the corrected flow rate from the sampler flow indicator, illustrated in Figure A.3(c), is addressed in Section 10.1.

Note: The following calibration procedure applies to a conventional orifice-type flow transfer standard and an orifice-type flow indicator in the sampler (most common types). For samplers using a pressure recorder having a square-root scale, 3 other acceptable calibration procedures are provided in Reference [12]. Other types of transfer standards may be used

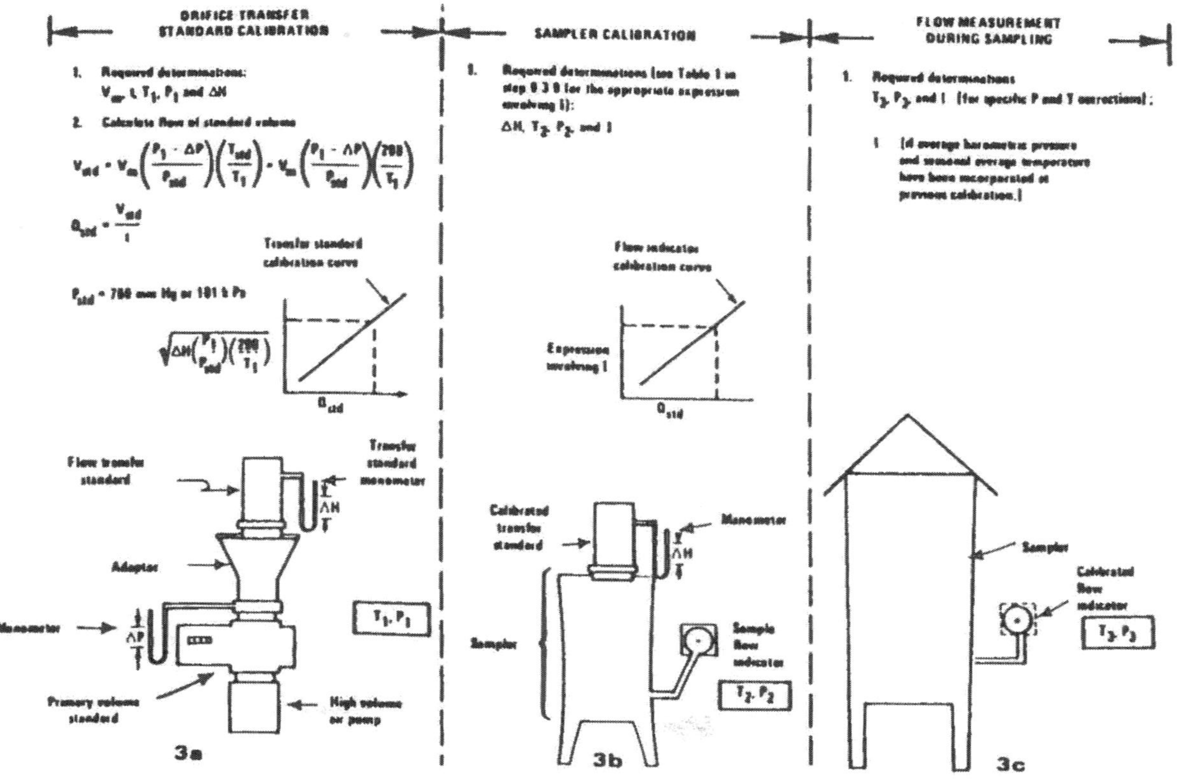

Figure A.3 Illustration of the three steps in the flow measurement process.

if the manufacturer or user provides an appropriately modified calibration procedure that has been approved by EPA.

9.2 *Certification of the flow rate transfer standard.*

9.2.1 *Equipment required:* Positive displacement standard volume meter traceable to the National Bureau of Standards (such as a Roots meter or equivalent), stop-watch, manometer, thermometer, and barometer.

9.2.2 Connect the flow rate transfer standard to the inlet of the standard volume meter. Connect the manometer to measure the pressure at the inlet of the standard volume meter. Connect the orifice manometer to the pressure tap on the transfer standard. Connect a high-volume air pump (such as a high-volume sampler blower) to the outlet side of the standard volume meter. See Figure A.3(a).

9.2.3 Check for leaks by temporarily clamping both manometer lines (to avoid fluid loss) and blocking the orifice with a large-diameter rubber stopper, wide cellophane tape, or other suitable means. Start the high-volume air pump and note any change in the standard volume meter reading. The reading should remain constant. If the reading changes, locate any leaks by listening for a whistling sound and/or retightening all connections, making sure that all gaskets are properly installed.

9.2.4 After satisfactorily completing the leak check as described above, unclamp both manometer lines and zero both manometers.

9.2.5 Achieve the appropriate flow rate through the system, either by means of the variable flow resistance in the transfer standard or by varying the voltage to the air pump. (Use of resistance plates as shown in Figure A.1(a) is discouraged because the above leak check must be repeated each time a new resistance plate is installed.) At least five different but constant flow rates, evenly distributed, with at least three in the specified flow rate interval (1.1 to 1.7 m^3/min [39–60 ft^3/min]), are required.

9.2.6 Measure and record the certification data on a form similar to the one illustrated in Figure A.4 according to the following steps.

9.2.7 Observe the barometric pressure and record as P_1 (item 8 in Figure A.4).

9.2.8 Read the ambient temperature in the vicinity of the standard volume meter and record it as T_1 (item 9 in Figure A.4).

9.2.9 Start the blower motor, adjust the flow, and allow the system to run for at least 1 min for a constant motor speed to be attained.

9.2.10 Observe the standard volume meter reading and simultaneously start a stopwatch. Record the initial meter reading (V_i) in column 1 of Figure A.4.

9.2.11 Maintain this constant flow rate until at least 3 m^3 of air have passed through the standard volume meter. Record the standard volume meter inlet pressure manometer reading as ΔP (column 5 in Figure A.4), and the orifice manometer reading as ΔH (column 7 in Figure A.4). Be sure to indicate the correct units of measurement.

Run No.	(1) Meter reading start V_i (m^3)	(2) Meter reading stop V_f (m^3)	(3) Sampling time t (min)	(4) Volume measured V_m (m^3)	(5) Differential pressure (at inlet to volume meter) ΔP (mm Hg or kPa)	(6) (X) Flow rate Q_{std} (std m^3/min)	(7) Pressure drop across orifice ΔH ☐ (in) or ☐ (cm) of water	(7a) ($\hat{Y}$) $\sqrt{\Delta H \left(\frac{P_1}{P_{std}}\right)\left(\frac{298}{T_1}\right)}$
1								
2								
3								
4								
5								
6								

RECORDED CALIBRATION DATA

Standard volume meter no. ______________________

Transfer standard type: ☐ orifice ☐ other

Model No. ____________ Serial No. ____________

(8) P_1: ____________ mm Hg (or kPa) (10) P_{std}: 760 mm Hg (or 101 kPa)

(9) T_1: ____________ K (11) T_{std}: 298 K

Calibration performed by: ______________________

Date: ____________

CALCULATION EQUATIONS

(1) $V_m = V_f - V_i$

(2) $V_{std} = V_m \left(\frac{P_1 - \Delta P}{P_{std}}\right)\left(\frac{T_{std}}{T_1}\right)$

(3) $Q_{std} = \frac{V_{std}}{t}$

LEAST SQUARES CALCULATIONS

Linear ($Y = mX + b$) regression equation of $Y = \sqrt{\Delta H(P_1/P_{std})(298/T_1)}$ on $X = Q_{std}$ for Orifice Calibration Unit (i.e., $\sqrt{\Delta H(P_1/P_{std})(298/T_1)} = mQ_{std} + b$)

Slope (m) = ____________ Intercept (b) = ____________ Correlation coefficient (r) = ____________

To use for subsequent calibration: $X = \frac{1}{m}(Y-b)$; $Q_{std} = \frac{1}{m}\left(\sqrt{\Delta H\left(\frac{P_2}{P_{std}}\right)\left(\frac{298}{T_2}\right)} - b\right)$

Figure A.4 Example of orifice transfer standard certification worksheet.

9.2.12 After at least 3 m^3 of air have passed through the system, observe the standard volume meter reading while simultaneously stopping the stopwatch. Record the final meter reading (V_f) in column 2 and the elapsed time (t) in column 3 of Figure A.4.

9.2.13 Calculate the volume measured by the standard volume meter at meter conditions of temperature and pressures as $V_m = V_f - V_1$. Record in column 4 of Figure A.4.

9.2.14 Correct this volume to standard volume (std m^3) as follows:

$$V_{std} = V_m \frac{P_1 - \Delta P}{P_{std}} \frac{T_{std}}{T_1}$$

where:

V_{std} = standard volume, std m^3
V_m = actual volume measured by the standard volume meter
P_1 = barometric pressure during calibration, mm Hg or kPa
Δ_P = differential pressure at inlet to volume meter, mm Hg or kPa
P_{std} = 760 mm Hg or 101 kPa
T_{std} = 298 K
T_1 = ambient temperature during calibration, K

Calculate the standard flow rate (std m^3/min) as follows:

$$Q_{std} = \frac{V_{std}}{t}$$

where:

Q_{std} = standard volumetric flow rate, std m^3/min
t = elapsed time, minutes

Record Q_{std} to the nearest 0.01 std m^3/min in column 6 of Figure A.4.

9.2.15 Repeat steps 9.2.9 through 9.2.14 for at least four additional constant flow rates, evenly spaced over the approximate range of 1.0 to 1.8 std m^3/min (35–64 ft^3/min).

9.2.16 For each flow, compute

$$\sqrt{\Delta H (P_1/P_{std})(298/T_1)}$$

(column 7a of Figure A.4) and plot these values against Q_{std} as shown in Figure A.3(a). Be sure to use consistent units (mm Hg or kPa) for barometric pressure. Draw the orifice transfer standard certification curve or calculate the linear least squares slope (m) and intercept (b) of the certification curve:

$$\sqrt{\Delta H(P_1/P_{std})(298/T_1)}$$

$= mQ_{std} + b$. See Figures A.3 and A.4. A certification graph should be readable to 0.02 std m³/min.

9.2.17 Recalibrate the transfer standard annually or as required by applicable quality control procedures (see Reference [2]).

9.3 *Calibration of sampler flow indicator.*

Note: For samplers equipped with a flow controlling device, the flow controller must be disabled to allow flow changes during calibration of the sampler's flow indicator, or the alternate calibration of the flow controller given in 9.4 may be used. For samplers using an orifice-type flow indicator downstream of the motor, do not vary the flow rate by adjusting the voltage or power supplied to the sampler.

9.3.1 A form similar to the one illustrated in Figure A.5 should be used to record the calibration data.

9.3.2 Connect the transfer standard to the inlet of the sampler. Connect the orifice manometer to the orifice pressure tap, as illustrated in Figure 3(b). Make sure there are no leaks between the orifice unit and the sampler.

9.3.3 Operate the sampler for at least 5 minutes to establish thermal equilibrium prior to the calibration.

9.3.4 Measure and record the ambient temperature, T_2, and the barometric pressure, P_2, during calibration.

9.3.5 Adjust the variable resistance or, if applicable, insert the appropriate resistance plate (or no plate) to achieve the desired flow rate.

9.3.6 Let the sampler run for at least 2 min to re-establish the run-temperature conditions. Read and record the pressure drop across the orifice (ΔH) and the sampler flow rate indication (I) in the appropriate columns of Figure A.5.

9.3.7 Calculate $\sqrt{\Delta H\,(P_2/P_{std})(298/T_2)}$ and determine the flow rate at standard conditions (Q_{std}) either graphically from the certification curve or by calculating Q_{std} from the least square slope and intercept of the transfer standard's transposed certification curve:

$$Q_{std} = 1/m \sqrt{\Delta H(P_2/P_{std})\,(298/T_2)} - b$$

Record the value of Q_{std} on Figure A.5.

9.3.8 Repeat steps 9.3.5, 9.3.6, and 9.3.7 for several additional flow rates distributed over a range that includes 1.1 to 1.7 std m³/min.

9.3.9 Determine the calibration curve by plotting values of the appropriate expression involving I, selected from Table A.1, against Q_{std}. The choice of expression from Table A.1 depends on the flow rate measurement device used (see Section 7.4.1) and also on whether the calibration curve is to incorporate geographic average barometric pressure (P_a) and seasonal

HIGH-VOLUME AIR SAMPLER CALIBRATION WORKSHEET

Site Location: ______________________

Date: ____________ Barometric Pressure, P_2 mm Hg (or kPa) ____________

Calibrated By: ____________ Temperature, T_2 (K) ____________

Sampler No. ____________ Serial No. ____________

Transfer std. type: ____________ Serial No. ____________

P_{std} = 760 mm Hg (or 101 kPa)

Optional:

Average barometric pressure: P_a = ____________

Seasonal average temperature: T_a = ____________

No.	ΔH Pressure drop across orifice ☐(in) or ☐(cm) of water	$\sqrt{\Delta H\left(\frac{P_2}{P_{std}}\right)\left(\frac{298}{T_2}\right)}$	(X) Q_{std} (from orifice certification) std m³/min	I Sampler flow rate indication (arbitrary)	(Y) For specific pressure and temperature corrections (see Table 1): ☐ I or ☐ $\sqrt{I\left(\frac{P_2}{P_{std}}\right)\left(\frac{298}{T_2}\right)}$ or ☐ $I\sqrt{\left(\frac{P_2}{P_{std}}\right)\left(\frac{298}{T_2}\right)}$	For incorporation of average pressure and seasonal average temperature (see Table 1): ☐ I or ☐ $\sqrt{I\left(\frac{P_2}{P_a}\right)\left(\frac{T_a}{T_2}\right)}$ or ☐ $I\sqrt{\left(\frac{P_2}{P_a}\right)\left(\frac{T_a}{T_2}\right)}$
1						
2						
3						
4						
5						
6						

LEAST SQUARES CALCULATIONS

Linear regression of Y on X: Y = mX + b; Y = appropriate expression from Table 1; X = Q_{std}.

Slope (m) = ____ Intercept (b) = ____ Correlation Coeff. (r) = ________

To determine subsequent flow rate during use: $X = \frac{1}{m}(Y-b)$; $Q_{std} = \frac{1}{m}$([appropriate expression from Table 2] - b)

Figure A.5 Example of high-volume air sampler calibration worksheet.

TABLE A.1. Expressions for Plotting Sampler Calibration Curves.

Type of Sampler Flow Rate Measuring Device	Expression: For Actual Pressure and Temperature Corrections	Expression: For Incorporation of Geographic Average Pressure and Seasonal Average Temperature
Mass flowmeter	I	I
Orifice and pressure indicator	$\sqrt{\left(\frac{P_s}{P_{std}}\right)\left(\frac{298}{T_2}\right)}$	$\sqrt{\left(\frac{P_s}{P_a}\right)\left(\frac{T_a}{T_2}\right)}$
Rotameter, or orifice and pressure recorder having square root scale*	$\sqrt{\left(\frac{P_2}{P_{std}}\right)\left(\frac{298}{T_2}\right)}$	$\sqrt{\left(\frac{P_2}{P_a}\right)\left(\frac{T_a}{T_2}\right)}$

*This scale is recognizable by its nonuniform divisions and is most commonly available for high-volume samplers.

average temperature (T_a) for the site to approximate actual pressure and temperature. Where P_a and T_a can be determined for a site for a seasonal period such that the actual barometric pressure and temperature at the site do not vary by more than ± 60 mm Hg (8 kPa) from P_a or ± 15°C from T_a, respectively, then using P_a and T_a avoids the need for subsequent pressure and temperature calculation when the sampler is used. The geographic average barometer pressure (P_a) may be estimated from an altitude-pressure table or by making an (approximate) elevation correction of -26 mm Hg (-3.46 kPa) for each 305 m (1,000 ft) above sea level (760 mm Hg or 101 kPa). The seasonal average temperature (T_a) may be estimated from weather station or other records. Be sure to use consistent units (mm Hg or kPa) for barometric pressure.

9.3.10 Draw the sampler calibration curve or calculate the linear least squares slope (m), intercept (b), and correlation coefficient of the calibration curve: [Expression from Table A.1] $= mQ_{std} + b$. See Figures A.3 and A.5. Calibration curves should be readable to 0.02 std m^3/min.

9.3.11 For a sampler equipped with a flow controller, the flow controlling mechanism should be re-enabled and set to a flow near the lower flow limit to allow maximum control range. The sample flow rate should be verified at this time with a clean filter installed. Then add two or more filters to the sampler to see if the flow controller maintains a constant flow; this is particularly important at high altitudes where the range of the flow controller may be reduced.

9.4 Alternate calibration of flow-controlled samplers. A flow-controlled sampler may be calibrated solely at its controlled flow rate, provided that previous operating history of the sampler demonstrates that the flow rate is stable and reliable. In this case, the flow indicator may remain uncalibrated but should be used to indicate any relative change between initial and final flows, and the sampler should be recalibrated more often to minimize potential loss of samples because of controller malfunction.

9.4.1 Set the flow controller for a flow near the lower limit of the flow range to allow maximum control range.

9.4.2 Install a clean filter in the sampler and carry out steps 9.3.2, 9.3.3, 9.3.4, 9.3.6, and 9.3.7.

9.4.3 Following calibration, add one or two additional clean filters to the sampler, reconnect the transfer standard, and operate the sampler to verify that the controller maintains the same calibrated flow rate; that is particularly important at high altitudes where the flow control range may be reduced.

10.0 *Calculations of TSP Concentration.*

10.1 Determine the average sampler flow rate during the sampling period according to either 10.1.1 or 10.1.2 below.

10.1.1 For a sampler without a continuous flow recorder, determine the appropriate expression to be used from Table A.2 corresponding to the one from Table A.1 used in step 9.3.9. Using this appropriate expression, determine Q_{std} for the initial flow rate from the sampler calibration curve, either graphically or from the transposed regression equation:

$$Q_{std} = 1/m \ [(\text{Appropriate expression from Table A.2} - b)]$$

Similarly, determine Q_{std} from the final flow reading, and calculate the average flow Q_{std} as one-half the sum of the initial and final flow rates.

10.1.2 For a sampler with a continuous flow recorder, determine the average flow rate device reading, I, for the period. Determine the ap-

TABLE A.2. **Expressions for Determining Flow Rate during Sampler Operation.**

Type of Sampler Flow Rate Measuring Device	Expression: For Actual Pressure and Temperature Corrections	Expression: For Use when Geographic Average Pressure and Seasonal Average Temperature Have Been Incorporated into the Sampler Calibration
Mass flowmeter	I	I
Orifice and pressure indicator	$\sqrt{\left(\frac{P_3}{P_{std}}\right)\left(\frac{298}{T_3}\right)}$	I
Rotameter, or orifice and pressure recorder having square root scale*	$\sqrt{\left(\frac{P_3}{P_{std}}\right)\left(\frac{298}{T_3}\right)}$	I

*This scale is recognizable by its nonuniform divisions and is most commonly available for high-volume samplers.

propriate expression from Table A.2 corresponding to the one from Table A.1 used in step 9.3.9. Then using this expression and the average flow rate reading, determine Q_{std} from the sampler calibration curve, either graphically or from the transposed regression equation:

$$Q_{std} = 1/m\ [(\text{Appropriate expression from Table A.2} - b)]$$

If the trace shows substantial flow change during the sampling period, greater accuracy may be achieved by dividing the sampling period into intervals and calculating an average reading before determining Q_{std}.

10.2 Calculate the total air volume sampled as:

$$V - Q_{std} \times t$$

where:

V = total air volume sampled, in standard volume units, std m³/min
Q_{std} = average standard flow rate, std m³/min
t = sampling time, min

10.3 Calculate and report the particulate matter concentration as:

$$TSP = \frac{(W_f - W_i) \times 10^6}{V}$$

where:

TSP = mass concentration of total suspended particulate matter, μg/std m³
W_i = initial weight of clean filter, g
W_f = final weight of exposed filter, g
V = air volume sampled, converted to standard conditions, std m³
10^6 = conversion of g to μg

10.4 If desired, the actual particulate matter concentration can be calculated as follows:

$$(TSP)_a = TSP(P_3/P_{std})(298/T_3)$$

where:

$(TSP)_a$ = actual concentration at field conditions, μg/m³
TSP = concentration at standard conditions, μ/std m³
P_3 = average barometric pressure during sampling period, mm Hg
P_{std} = 760 mm Hg (or 101 kPa)
T_3 = average ambient temperature during sampling period, K

11.0 *References.*

1 *Quality Assurance Handbook for Air Pollution Measurement Systems, Volume I, Principles*. EPA-600 /9-76-005, U.S. Environmental Protection Agency, Research Triangle Park, NC 27711, 1976.

2 *Quality Assurance Handbook for Air Pollution Measurement Systems, Volume II, Ambient Air Specific Methods*. EPA-600 /4-77-027a, U.S. Environmental Protection Agency, Research Triangle Park, NC 27711, 1977.

3 Wedding, J. B., A. R. McFarland and J. E. Cernak. "Large Particle Collection Characteristics of Ambient Aerosol Samplers," *Environ. Sci. Technol*. 11:387-390, 1977.

4 McKee, H. C. et al. "Collaborative Testing of Methods to Measure Air Pollutants, I. The High-Volume Method for Suspended Particulate Matter," *J. Air Poll. Cont. Assoc*., 22(342), 1972.

5 Clement, R. E. and F. W. Karasek. "Sample Composition Changes in Sampling and Analysis of Organic Compounds in Aerosols," *The Intern. J. Environ. Anal.* Chem., 7:109, 1979.

6 Lee, R. E., Jr. and J. Wagman. "A Sampling Anomaly in the Determination of Atmosphere Sulfuric Concentration," *Am. Ind. Hygiene Assoc. J*., 27:266, 1966.

7 Appel, B. R. et al. "Interference Effects in Sampling Particulate Nitrate in Ambient Air," *Atmospheric Environment*, 13:319, 1979.

8 Tierney, G. P. and W. D. Conner. "Hygroscopic Effects on Weight Determination of Particulates Collected on Glass-Fiber Filters," *Am. Ind. Hygiene Assoc. J*., 28:363, 1967.

9 Chahal, H.S. and D. J. Romano. "High-Volume Sampling Effect of Windborner" Particulate Matter Deposited during Idle Periods," *J. Air Poll. Cont. Assoc*., 26(885), 1976.

10 Patterson, R. K. "Aerosol Contamination from High-Volume Sampler Exhaust," *J. Air Poll. Cont. Assoc*., Vol. 30(169), 1980.

11 EPA Test Procedures for Determining pH and Integrity of High-Volume Air Filters. QAD/M-80.01. Available from the Methods Standardization Branch, Quality Assurance Division, Environmental Monitoring Systems Laboratory (MD-77), U.S. Environmental Protection Agency, Research Triangle Park, NC 27711, 1980.

12 Smith, F., P. S. Wohlschlegel, R. S. C. Rogers and D. J. Mulligan. "Investigation of Flow Rate Calibration Procedures Associated with the High-Volume Method for Determination of Suspended Particulates," EPA-600/4-78-047, U.S. Environmental Protection Agency, Research Triangle Park, NC, June 1978.

Reference Method for the Determination of Lead in Suspended Particulate Matter Collected from Ambient Air

1. *Principle and applicability.*

1.1 Ambient air suspended particulate matter is collected on a glass-fiber filter for 24 hours using a high volume air sampler.

1.2 Lead in the particulate matter is solubilized by extraction with nitric acid (HNO_3), facilitated by heat or by a mixture of HNO_3 and hydrochloric acid (HCl) facilitated by ultrasonication.

1.3 The lead content of the sample is analyzed by atomic absorption spectrometry using an air-acetylene flame, the 283.3 or 217.0 nm lead absorption line, and the optimum instrumental conditions recommended by the manufacturer.

1.4 The ultrasonication extraction with HNO_3/HCl will extract metals other than lead from ambient particulate matter.

2. *Range, sensitivity, and lower detectable limit.* The values given below are typical of the method's capabilities. Absolute values will vary for individual situations depending on the type of instrument used, the lead line, and operating conditions.

2.1 *Range.* The typical range of the method is 0.07 to 7.5 μg Pb/m^3 assuming an upper linear range of analysis of 15/μg·ml and an air volume of 2,400 m^3.

2.2 *Sensitivity.* Typical sensitivities for a 1 percent change in absorption (0.0044 absorbance units) are 0.2 and 0.5 μg Pb/ml for the 217.0 and 283.3 nm lines, respectively.

2.3 Lower detectable limit (LDL). A typical LDL is 0.07 μg Pb/m^3. The above value was calculated by doubling the between-laboratory standard deviation obtained for the lowest measurable lead concentration in a collaborative test of the method [15]. An air volume of 2,400 m^3 was assumed.

3. *Interferences.* Two types of interferences are possible: chemical and light scattering.

3.1 *Chemical.* Reports on the absence [1–5] of chemical interferences far outweigh those reporting their presence [6], therefore, no correction for chemical interferences is given here. If the analyst suspects that the sample matrix is causing a chemical interference, the interference can be verified and corrected for by carrying out the analysis with and without the method of standard additions [7].

3.2 *Light scattering*. Nonatomic absorption or light scattering, produced by high concentrations of dissolved solids in the sample, can produce a significant interference, especially at low lead concentrations [2]. The interference is greater at the 217.0 nm line than at the 283.3 nm line. No interference was observed using the 283.3 nm line with a similar method [1].

Light scattering interferences can, however, be corrected for instrumentally. Since the dissolved solids can vary depending on the origin of the sample, the correction may be necessary, especially when using the 217.0 nm line. Dual beam instruments with a continuum source give the most accurate correction. A less accurate correction can be obtained by using a nonabsorbing lead line that is near the lead analytical line. Information on use of these correction techniques can be obtained from instrument manufacturers' manuals.

If instrumental correction is not feasible, the interference can be eliminated by use of the ammonium pyrrolidinecarbodithioatemethylisobutyl ketone, chelation-solvent extraction technique of sample preparation [8].

4. *Precision and bias.*

4.1 The high-volume sampling procedure used to collect ambient air particulate matter has a between-laboratory relative standard deviation of 3.7 percent over the range 80 to 125 $\mu g/m^3$ [9]. The combined extraction-analysis procedure has an average within-laboratory relative standard deviation of 5 to 6 percent over the range 1.5 to 15 μg Pb/ml, and an average between laboratory relative standard deviation of 7 to 9 percent over the same range. These values include use of either extraction procedure.

4.2 Single laboratory experiments and collaborative testing indicate that there is no significant difference in lead recovery between the hot and ultrasonic extraction procedures [15].

5. *Apparatus.*

5.1 Sampling.

5.1.1 High-volume sampler. Use and calibrate the sampler as described in Reference [10].

5.2 Analysis.

5.2.1 Atomic absorption spectrophotometer. Equipped with lead hollow cathode or electrodeless discharge lamp.

5.2.1.1 Acetylene. The grade recommended by the instrument manufacturer should be used. Change cylinder when pressure drops below 50–100 psig.

5.2.1.2 Air. Filtered to remove particulate, oil, and water.

5.2.2 Glassware. Class A borosilicate glassware should be used throughout the analysis.

5.2.2.1 Beakers, 30 and 150 ml graduated, Pyrex.

5.2.2.2 Volumetric flasks, 100-ml.

5.2.2.3 Pipettes. To deliver 50, 30, 15, 8, 4, 2, 1 ml.

5.2.2.4 Cleaning. All glassware should be scrupulously cleaned. The following procedure is suggested. Wash with laboratory detergent, rinse, soak for 4 hours in 20 percent (w/w) HNO_3, rinse 3 times with distilled deionized water, and dry in a dust free manner.

5.2.3 Hot plate.

5.2.4 Ultrasonication water bath, unheated. Commercially available laboratory ultrasonic cleaning baths of 450 watts or higher "cleaning power," i.e., actual ultrasonic power output to the bath has been found satisfactory.

5.2.5 Template. To aid in sectioning the glass-fiber filter. See Figure B.1 for dimensions.

5.2.6 Pizza cutter. Thin wheel. Thickness <1 mm.

5.2.7 Watch glass.

5.2.8 Polyethylene bottles. For storage of samples. Linear polyethylene gives better storage stability than other polyethylenes and is preferred.

5.2.9 Parafilm "M"[11] American Can Co., Marathon Products, Nennah, Wis., or equivalent.

6. *Reagents.*

6.1 Sampling.

6.1.1 Glass fiber filters. The specifications given below are intended to aid the user in obtaining high quality filters with reproducible properties. These specifications have been met by EPA contractors.

6.1.1.1 Lead content. The absolute lead content of filters is not critical, but low values are, of course, desirable. EPA typically obtains filters with a lead content of $<75\ \mu g$/filter.

It is important that the variation in lead content from filter to filter, within a given batch, be small.

6.1.1.2 Testing.

6.1.1.2.1 For large batches of filters (>500 filters) select at random 20 to 30 filters from a given batch. For small batches (<500 filters) a lesser number of filters may be taken. Cut one 3/4" × 8" strip from each filter anywhere in the filter. Analyze all strips, separately, according to the directions in sections 7 and 8.

6.1.1.2.2 Calculate the total lead in each filter as

[11]Mention of commercial products does not imply endorsement by the U.S. Environmental Protection Agency.

$$F_b = \mu\text{g Pb/ml} \times \frac{100\text{ ml}}{\text{strip}} \times \frac{12\text{ strips}}{\text{filter}}$$

where:

F_b = amount of lead per 72 square inches of filter, μg

6.1.1.2.3 Calculate the mean, F_b of the values and the relative standard deviation (standard deviation/mean × 100). If the relative standard deviation is high enough so that, in the analyst's opinion, subtraction of F_b (section 10.3) may result in a significant error in the μg Pb/m^2 the batch should be rejected.

6.1.1.2.4 For acceptable batches, use the value of F_b to correct all lead analyses (section 10.3) of particulate matter collected using that batch of filters. If the analyses are below the LDL (section 2.3) no correction is necessary.

6.2 Analysis.

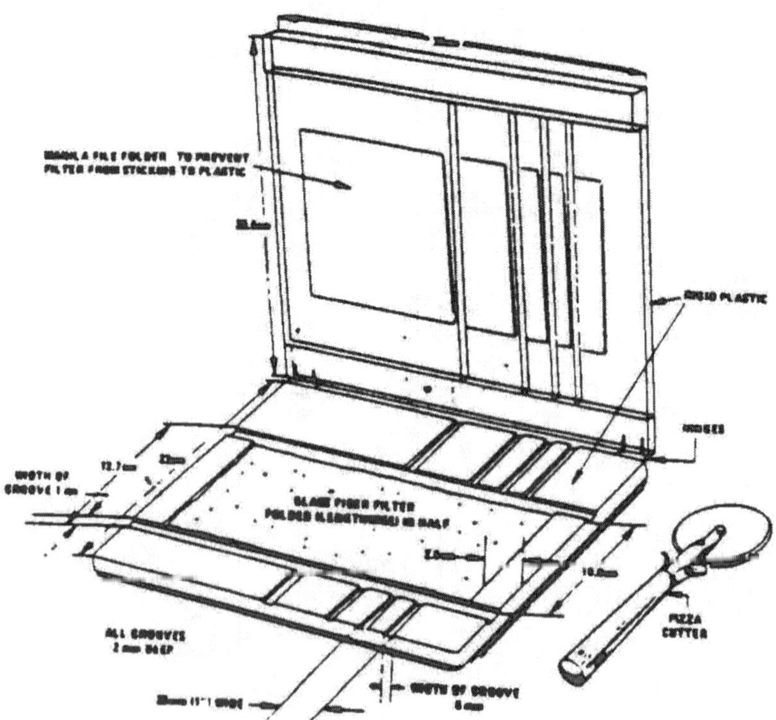

Figure B.1 Rules and regulations.

6.2.1 Concentrated (15.6 *M*) HNO_3. ACS reagent grade HNO_3 and commercially available redistilled HNO_3 have been found to have sufficiently low lead concentrations.

6.2.2 Concentrated (11.7 *M*) HCl. ACS reagent grade.

6.2.3 Distilled-deionized water (D.I. water)

6.2.4 3 *M* HNO_3. This solution is used in the hot extraction procedure. To prepare, add 192 ml of concentrated HNO_3 to D.I. water in a 1 l volumetric flask. Shake well, cool, and dilute to volume with D.I. water. *Caution:* Nitric acid fumes are toxic. Prepare in a well ventilated fume hood.

6.2.5 0.45 *M* HNO_3. This solution is used as the matrix for calibration standards when using the hot extraction procedure. To prepare, add 29 ml of concentrated HNO_3 to D.I. water in a 1 l volumetric flask. Shake well, cool, and dilute to volume with D.I. water.

6.2.6 2.6 *M* HNO_3 + 0 to 0.9 *M* HCl. This solution is used in the ultrasonic extraction procedure. The concentration of HCl can be varied from 0 to 0.9 *M*. Directions are given for preparation of 2.6 *M* HNO_3 + 0.9 *M* HCl solution. Place 167 ml of concentrated HNO_3 into a 1 l volumetric flask and add 77 ml of concentrated HCl. Stir 4 to 6 hours, dilute to nearly 1 l with D.I. water, cool to room temperature, and dilute to 1 l.

6.2.7 0.40 *M* HNO_3 × X *M* HCl. This solution is used as the matrix for calibration standards when using the ultrasonic extraction procedure. To prepare, add 26 ml of concentrated HNO_3 plus the ml of HCl required, to a 1 l volumetric flask. Dilute to nearly 1 l with D.I. water, cool to room temperature, and dilute to 1 l. The amount of HCl required can be determined from the following equation:

$$y = \frac{77 \text{ ml} \times 0.15x}{0.9\ M}$$

where:

y = ml of concentrated HCl required
x = molarity of HCl in 6.2.6
0.15 = dilution factor in 7.2.2

6.2.8 Lead nitrate, $Pb(NO_3)_3$. ACS reagent grade, purity 99.0 percent. Heat for 4 hours at 120°C and cool in a desiccator.

6.3 Calibration standards.

6.3.1 Master standard, 1000 μg Pb/ml in HNO_3. Dissolve 1.598 g of $Pb(NO_3)_3$ in 0.45 *M* HNO_3 contained in a 1 l volumetric flask and dilute to volume with 0.45 *M* HNO_3.

6.3.2 Master standard, 1000 μg Pb/ml in HNO_3/HCl. Prepare as in 6.3.1 except use the HNO_3 /HCl solution in 6.2.7.

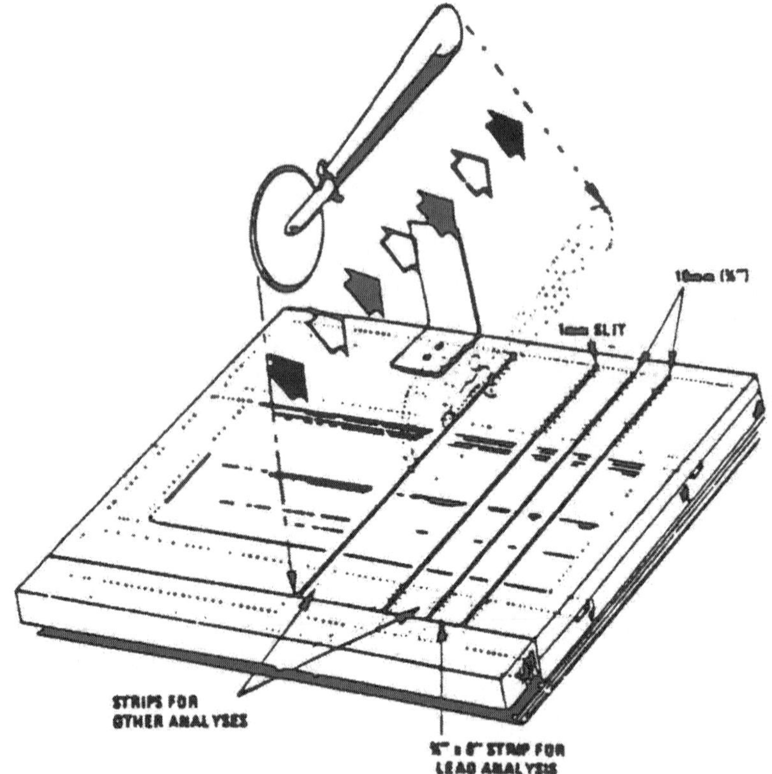

Figure B.2 Rules and regulations.

Store standards in a polyethylene bottle. Commercially available certified lead standard solutions may also be used.

7. *Procedure.*

7.1 Sampling. Collect samples for 24 hours using the procedure described in Reference [10] with glass-fiber filters meeting the specifications in 6.1.1. Transport collected samples to the laboratory taking care to minimize contamination and a loss of sample [17].

7.2 Sample preparation.

7.2.1 Hot extraction procedure.

7.2.1.1 Cut a 3/4″ × 8″ strip from the exposed filter using a template and a pizza cutter as described in Figures B.1 and B.2. Other cutting procedures may be used.

Lead in ambient particulate matter collected on glass fiber filters has been shown to be uniformly distributed across the filter [1,3,11] suggesting that the position of the strip is unimportant. However, another study [12] has shown that when sampling near a roadway lead is not uniformly distributed across the filter. The nonuniformity has been attributed to large variations in particle size [16]. Therefore, when sampling near a roadway additional strips at different positions within the filter should be analyzed.

7.2.1.2 Fold the strip in half twice and place in a 150-ml beaker. Add 15 ml of 3 *M* HNO_3 to cover the sample. The acid should completely cover the sample. Cover the beaker with a watch glass.

7.2.1.3 Place beaker on the hot-plate, contained in a fume hood, and boil gently for 30 min. Do not let the sample evaporate to dryness. *Caution:* Nitric acid fumes are toxic.

7.2.1.4 Remove beaker from hot plate and cool to near room temperature.

7.2.1.5 Quantitatively transfer the sample as follows:

7.2.1.5.1 Rinse watch glass and sides of beaker with D.I. water.

7.2.1.5.2 Decant extract and rinsings into a 100-ml volumetric flask.

7.2.1.5.3 Add D.I. water to 40 ml mark on beaker, cover with watch glass, and set aside for a minimum of 30 minutes. This is a critical step and cannot be omitted since it allows the HNO_3 trapped in the filter to diffuse into the rinse water.

7.2.1.5.4 Decant the water from the filter into the volumetric flask.

7.2.1.5.5 Rinse filter and beaker twice with D.I. water and add rinsings to volumetric flask until total volume is 80 to 85 ml.

7.2.1.5.6 Stopper flask and shake vigorously. Set aside for approximately 5 minutes or until foam has dissipated.

7.2.1.5.7 Bring solution to volume with D.I. water. Mix thoroughly.

7.2.1.5.8 Allow solution to settle for one hour before proceeding with analysis.

7.2.1.5.9 If sample is to be stored for subsequent analysis, transfer to a linear polyethylene bottle.

7.2.2 Ultrasonic extraction procedure.

7.2.2.1 Cut a 3/4″ × 8″ strip from the exposed filter as described in section 7.2.1.1.

7.2.2.2 Fold the strip in half twice and place in a 30-ml beaker. Add 15 ml of the NHO_3/HCl solution in 6.2.6. The acid should completely cover the sample. Cover the beaker with parafilm.

The parafilm should be placed over the beaker such that none of the parafilm is in contact with water in the ultrasonic bath. Otherwise, rinsing of the parafilm (section 7.2.2.4.1) may contaminate the sample.

7.2.2.3 Place the beaker in the ultrasonication bath and operate for 30 minutes.

7.2.2.4 Quantitatively transfer the sample as follows:

7.2.2.4.1 Rinse parafilm and sides for beaker with D.I. water.

7.2.2.4.2 Decant extract and rinsings into a 100 ml volumetric flask.

7.2.2.4.3 Add 20 ml D.I. water to cover the filter strip, cover with parafilm and set aside for a minimum of 30 minutes. This is a critical step and cannot be omitted. The sample is then processed as in sections 7.2.1.5.4 through 7.2.1.5.9.

Note: Samples prepared by the hot extraction procedure are now in 0.45 M HNO_3. Samples prepared by the ultrasonication procedure are in 0.40 M HNO_3 + X M HCl.

8. *Analysis.*

8.1 Set the wavelength of the monochromator at 283.3 or 217.0 nm. Set or align other instrumental operating conditions as recommended by the manufacturer.

8.2 The sample can be analyzed directly from the volumetric flask, or an appropriate amount of sample decanted into a sample analysis tube. In either case, care should be taken not to disturb the settled solids.

8.3 Aspirate samples, calibration standards and blanks (section 9.2) into the flame and record the equilibrium absorbance.

8.4 Determine the lead concentration in μg Pb/ml, from the calibration curve, section 9.3.

8.5 Samples that exceed the linear calibration range should be diluted with acid of the same concentration as the calibration standards and reanalyzed.

9. *Calibration.*

9.1 Working standard, 20 μg Pb/ml. Prepared by diluting 2.0 ml of the master standard (6.3.1 if the hot acid extraction was used or 6.3.2 if the ultrasonic extraction procedure was used) to 100 ml with acid of the same concentration as used in preparing the master standard.

9.2 Calibration standards. Prepare daily by diluting the working standard, with the same acid matrix, as indicated below. Other lead concentrations may be used.

Volume of 20 μg/ml Working Standard, ml	Final Volume, ml	Concentration, μg Pb/ml
0	100	0
1.0	200	0.1
2.0	200	0.2
3.0	100	0.4
4.0	100	0.8
5.0	100	1.6
15.0	100	3.0
30.0	100	6.0
50.0	100	10.0
100.0	100	30.0

9.3 Preparation of calibration curve. Since the working range of analysis will vary depending one which lead line is used and the type of instrument, no one set of instructions for preparation of a calibration curve can be given. Select standards (plus the reagent blank), in the same acid concentration as the samples, to cover the linear absorption range indicated by the instrument manufacturer. Measure the absorbance of the blank and standards as in section 8.0. Repeat until good agreement is obtained between replicates. Plot absorbance (*y*-axis) versus concentration in μg Pb/ml (*x*-axis). Draw (or compute) a straight line through the linear portion of the curve. Do not force the calibration curve through zero. Other calibration procedures may be used.

To determine stability of the calibration curve, remeasure—alternately—one of the following calibration standards for every 10th sample analyzed: concentration $\leq$ 1 μg Pb/ml; concentration $\leq$ 10 μg Pb/ml. If either standard deviates by more than 5 percent from the value predicted by the calibration curve, recalibrate and repeat the previous 10 analyses.

10. *Calculation.*

10.1 Measured air volume. Calculate the measured air volume as

$$V_m \frac{Q_i + Q_f}{2} \times T$$

where:

V_m = air volume sampled (uncorrected), m³
Q_1 = initial air flow rate, m³/min
Q_f = final air flow rate, m³/min
T = sampling time, min

The flow rates Q_i and Q_f should be corrected to the temperature and pressure conditions existing at the time of orifice calibration as directed in addendum B of Reference [10], before calculation V_m.

10.2 Air volume at STP. The measured air volume is corrected to reference conditions of 760 mm Hg and 25°C as follows. The units are standard cubic meters, sm³.

$$V_{STP} = V_m \times \frac{P_2 \times T_1}{P_1 \times T_2}$$

V_{STP} = sample volume, sm³, at 760 mm Hg and 298°K
V_m = measured volume from 10.1
P_2 = atmospheric pressure at time of orifice calibration, mm Hg
P_1 = 760 mm Hg
T_2 = atmospheric temperature at time of orifice calibration, °K
T_1 = 298°K

10.3 Lead concentration. Calculate lead concentration in the air sample.

$$C = \frac{(\mu g\ Pb/ml = 100\ ml/strip = 12\ strips/filter) - F_b}{V_{STP}}$$

where:

C = concentration, μg Pb/sm^3
μg Pb/ml = lead concentration determined from section 8
100 ml/strip = total sample volume
12 strips/filter = usable filter area, 7″ × 9″ /exposed area of one strip 3/4″ × 7″
F_b = lead concentration of blank filter, μg, from section 6.1.1.2.3
V_{STP} = air volume from 10.2

11. *Quality control.*

3/4″ × 8″ glass fiber filter strips containing 80 to 2000 μg Pb/strip (as lead salts) and blank strips with zero Pb content should be used to determine if the method—as being used—has any bias. Quality control charts should be established to monitor differences between measured and true values. The frequency of such checks will depend on the local quality control program.

To minimize the possibility of generating unreliable data, the user should follow practices established for assuring the quality of air pollution data [13] and take part in EPA's semiannual audit program for lead analyses.

12. *Trouble shooting.*

1. During extraction of lead by the hot extraction procedure, it is important to keep the sample covered so that corrosion products—formed on fume hood surfaces which may contain lead—are not deposited in the extract.
2. The sample acid concentration should minimize corrosion of the nebulizer. However, different nebulizers may require lower acid concentrations. Lower concentrations can be used provided samples and standards have the same acid concentration.
3. Ashing of particulate samples has been found, by EPA and contractor laboratories, to be unnecessary in lead analyses by atomic absorption. Therefore, this step was omitted from the method.
4. Filtration of extracted samples, to remove particulate matter, was specifically excluded from sample preparation, because some analysts have observed losses of lead due to filtration.
5. If suspended solids should clog the nebulizer during analysis of samples, centrifuge the sample to remove the solids.

13. *Authority.*

(Secs. 109 and 301(a), Clean Air Act as amended (42 U.S.C. 7409, 7601(a)).)

14. *References.*

1 Scott, D. R. et al. 1976. "Atomic Absorption and Optical Emission Analysis of NASN Atmospheric Particulate Samples for Lead." Envir. Sci. and Tech., 10:877–880.

2 Skogerboe, R. K. et al. 1976. "Monitoring for Lead in the Environment," pp. 57–66, Department of Chemistry, Colorado State University, Fort Collins, Colo. 90523. Submitted to National Science Foundation for publication.

3 Zdrojewski, A. et al. 1972. "The Accurate Measurement of Lead in Airborne Particulates," *Inter. J. Environ. Anal. Chem.* 2:63–77.

4 Slavin, W. 1968. *Atomic Absorption Spectroscopy*. Published by Interscience Company, New York, N.Y..

5 Kirkbright, G. F. and M. Sargent, 1974. *Atomic Absorption and Fluorescence Spectroscopy* . Published by Academic Press, New York, N.Y.

6 Burnham, C. D. et al., "Determination of Lead in Airborne Particulates in Chicago and Cook County, Ill, by Atomic Absorption Spectroscopy." *Envir. Sci. and Tech.*, 3: 472–475 (1969).

7 "Proposed Recommended Practices for Atomic Absorption Spectrometry," *ASTM Book of Standards*, part 30, pp. 1596–1608 (July 1973).

8 Koirttyohann, S. R. and J. W. Wen, 1973. "Critical Study of the APCD-MIBK Extraction System for Atomic Absorption." *Anal. Chem.*, 45:1966–1989.

9 "Collaborative Study of Reference Method for the Determination of Suspended Particulates in the Atmosphere (High Volume Method)," Obtainable from National Technical Information Service, Department of Commerce, Port Royal Road, Springfield, VA, 22151, as PB-205-891.

10 "Reference Method for the Determination of Suspended Particulates in the Atmosphere (High Volume Method)." Code of Federal Regulations, Title 40. Part 50, Appendix B, pp. 12–16 (July 1, 1975).

11 Dubois, L. et al., 1966. "The Metal Content of Urban Air," *JAPCA*, 16:77–78.

12 EPA Report No. 600/4-77-034. June 1977. *Los Angeles Catalyst Study Symposium*, p. 223.

13 1976. *Quality Assurance Handbook for Air Pollution Measurement System. Volume 1–Principles.* EPA-600/9-76-005.

14 Thompson, R. J. et al. 1970. "Analysis of Selected Elements in Atmospheric Particulate Matter by Atomic Absorption" *Atomic Absorption Newsletter*, 9(3).

15 To be published. EPA. QAB. EMSL. RTP. N.C. 2771.

16 Hirschler, D. A. et al. 1957. "Particulate Lead Compounds in Automobile Exhaust Gas," *Industrial and Engineering Chemistry.* 40:1131–1142.

17 *Quality Assurance Handbook for Air Pollution Measurement Systems. Volume II–Ambient Air Specific Methods.* EPA-600/4-77/027a, May 1977.

Index

Printed and bound by CPI Group (UK) Ltd, Croydon, CR0 4YY
22/10/2024
01777622-0004